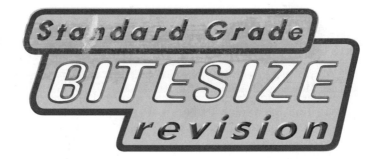

Standard Grade
BITESIZE
revision

Biology

Robert Dickson
Tony Aitken

Published by BBC Educational Publishing,
BBC White City, 201 Wood Lane, London W12 7TS
First published 2000, Reprinted 2001
© Robert Dickson/Tony Aitken/BBC Worldwide
(Educational Publishing), 2000

ISBN: 0 563 47490 4

Designed by Linda Reed and Associates
Illustrations by Tech Type
Printed in Great Britain by Bell & Bain Ltd., Glasgow

D0510339

Contents

About BITESIZE Biology for Standard Grade

This book is a revision guide to help you to do your best in your Standard Grade Biology exam. You can help yourself in three ways: work through the activities and questions in the book, watch the TV programmes or log on to the Internet to get more practice at the ideas you find difficult.

This book is a good basis for your revision, but do not forget to use your notes from school and any textbooks you have, too.

Read through this introduction carefully because it tells you how the book is laid out and how to use it. It tells you about the television programmes, and how to make the best use of them alongside this book; a little bit about topics covered by the programmes and some general help with exams and syllabi; a check list of the key things you should be able to do; and some tips to pick up those crucial extra marks in the exams.

About this book

This book covers the whole Standard Grade course, and is divided into the seven topics that you study.

KEY TO SYMBOLS

(?) Something to think about

(◎) An activity to do

(TV) A link to the video

(◉) A link to the website

Each topic begins with an introductory page to tell you what the topic is about. This summarises the main ideas in the topic. The next page is the FactZone, which lists key words and phrases that you need to know. The definitions are clear and concise, but rather than just skimming through them, you should look carefully at each definition and match it to the correct word.

The following pages give you more information, with activities to help you revise, questions to do and tips from the examiner.

There are practice questions after each topic, similar to the questions you might get in an exam, with answers at the back of the book.

The book has been designed to go alongside the BITESIZE television programmes and the BITESIZE website. You will see the TV symbol in the text (TV). This shows you where there is a link to the TV programme.

Using this book to revise

Planning your revision

Do you know when your exam is? How long have you got to revise? It is no good leaving revision until the night before the exam. The best way to revise is to break the subject up into BITESIZE chunks. That is why Standard Grade Biology is broken into topics and sub-topics. There are seven topics and 96 pages of biology to revise in this book. Use the contents page to plan your revision. You could write the date by each topic, to show when you will revise it.

Of course, you have other subjects to revise, too. It is often better to cover more than one subject in an evening. So how about planning all your revision by working out how much time you have before the exams start, and then sharing the days out amongst your subjects? Do not forget to leave some time to relax, too!

Revision tips

It is no good just sitting reading this book – to learn the material and understand it you need to be *active*. Here are some ideas to try:

- do each of the activities indicated by the ⑦ or ◉ symbols

- at the end of each double page, close the book and write down the key facts in your own words

- when there is a labelled diagram to learn, draw a copy of the diagram without the labels. Look at the labels in the book; close the book; label the diagram and then check how many were correct. 'Look, cover, write, check' is a good way of learning all sorts of things

- use information from the FactZone pages to make some flash cards: write a definition on one side of the card and the word on the other. Look at the word – can you write a definition? Look at the definition – which word is it?

- revise with a friend – flash cards are more fun with a partner.

Using the TV programmes

You will have to record the BITESIZE Standard Grade Biology programmes – unless you want to do all your revision in the middle of the night. The great thing about watching the recorded tape is that you can go back over the bits you do not understand as many times as you like! The whole programme is two hours long and covers all of the topics in the course, with lots of information.

It would be helpful if you have this BITESIZE Biology book with you as you watch the tape. Do not try to watch the whole tape at one time – watch one Bite and then work through that topic in the book. Do not be afraid to make plenty of notes in the margin – research has shown conclusively that you learn much better when you are actively involved in the process – making notes in your own words, summarising, identifying key information and so on.

The video has references to show where the book has more information. The book has TV icons to show there is a section of the programme to watch.

Standard Grade Biology

The course is decided by the Scottish Qualifications Authority and consists of seven topics, which are usually studied over two school sessions. There are three elements in the course: Knowledge and Understanding, Problem Solving, and Practical Abilities. The first two are assessed by an exam at the end of the

❗ The **REMEMBER!** paragraphs mention important points that you should note.

THE ONLINE SERVICE
◉ You can find extra support, tips and answers to your exam queries on the BITESIZE website. The address is www.bbc.co.uk/scotland/revision/

course, and count for 80% of your overall grade. The Practical Abilities are assessed during the course by your own school. This book is concerned with helping you with the exam. This is the part that earns most of the marks.

SQA exams

Everybody sits two papers – General and Credit – there is no Foundation level in Biology. Each paper is 90 minutes long and is exactly half Problem Solving (PS) and half Knowledge and Understanding (KU). The General paper is out of 100, whilst the Credit is out of 80. Your final grade will be the best grade you get in either paper so work for both papers unless your teacher has told you otherwise.

As everybody sits General and Credit papers, this book has not distinguished what material and questions are specific to Credit level. If you are opting out of the Credit paper, ask your teacher about what you don't need to revise.

The exam papers

The questions are usually in several parts. They often start with some information or a diagram. It is important that you read the information and look at the diagram to make sure you know what the question is about. Underline any important words in the information – they will help you when you come to write your answers.

Knowledge and Understanding

Most of these questions will test your recall of the facts you have learned during the course. They will ask you to answer in a variety of ways, by writing a word or short sentence, by ticking the correct box, or by underlining the correct answer from a range of choices. In every case be careful to read carefully what you are asked to do. Try to answer clearly, and use the proper biological words wherever possible – the FactZones will be a great help here. You should also try to spell words correctly, but don't worry too much if you forget the exact spelling – just make sure that the examiner can clearly identify the word you are trying to write – marks are almost never deducted for bad spelling.

Problem Solving

Don't let the name of this element worry you – you are not expected to be able to solve all the world's problems. In fact, there is a very specific list of skills that you require, and if you can achieve these, then you are sure of a very good grade. You need to be able to both draw and interpret graphs, charts, and tables; do simple calculations of averages, percentages and ratios; comment on experimental methods and, in particular, how to get valid and reliable results, and to draw a reasonable conclusion from a set of results.

One or two points might be helpful about drawing graphs as marks are lost quite unnecessarily here. When you draw a graph make sure that each axis has a proper scale with each interval equally spaced. Each axis needs a label and units. Many candidates miss out the zero on their scales — put it in, or risk losing a mark. Use crosses to plot points and be accurate — there is very little tolerance, and even one point out of place will lose the mark. Join the points with a ruler and make sure that the line goes through every cross. Finally, if the origin (0,0) is not given as one of the points on the graph, then you must not continue your line down to it. Your line must go from the first point to the last and not anywhere outside that range.

Bar charts are often very badly done. The top of the bar must be exactly along the correct value. If any space can be seen, then you will lose a mark. This means that wobbly lines drawn at the top of the bar without a ruler very often fail to get the mark.

Picking up marks in exams!

Follow these tips to make sure you get all the marks you deserve. The examiners cannot read your mind – they can only give you marks for what you actually put down on paper.

- Read the questions carefully – they contain the clues to the answers.

- Try to write something for every question. You'd be amazed how often you get the mark, even when you are not confident that your answer is correct. A blank space never ever gets a mark!

- Experiments – you repeat an experiment to make it more reliable. Don't even consider the word 'accurate' and remember that 'so that you can take an average' is not the answer.

- If you are asked to improve the apparatus, then check for leaks, check corks are properly fitted and that thermometers are in the right place — most errors are of this type.

By buying this book and reading this introduction, you have already shown that you are determined to do as well as you can. Good preparation and careful revision is the key to success in any exam.

Good luck!

The biosphere

This topic is about:

- **investigating an ecosystem**
  sampling methods, measuring abiotic factors

- **how it works**
  food chains, food webs, pyramids of numbers and biomass,
  population growth curves, competition, the need for recycling,
  the nitrogen cycle

- **control and management**
  sources of pollution, the effects of energy sources on the
  environment, control of pollution, indicator species, results of
  poor management, control of ecosystems by agriculture

Much of biology is about individual organisms and the cells and tissues that go to make them. This topic, however, is concerned with larger issues. Ecology is all about the interaction amongst individuals, species, populations, communities and their habitats to make fully functioning ecosystems.

In studying an ecosystem, we must first ask ourselves exactly what it consists of. What species are present, and in what numbers? This involves careful observation, counting, measuring and sampling. We will consider some of the methods used by ecologists in this work, and the ways in which they seek to minimise potential errors.

Once we are aware of what is present in the system, we can start to build our knowledge of how the jigsaw fits together. What eats what? What factors control and limit the distribution and size of each population? What happens to the important resources of the system? Even for the smallest of ecosystems, the answers to these questions are likely to be very complex, and so we must also include in the skills we require some ideas about ways in which we

can represent our findings so that others can understand and learn from them. Food web diagrams and pyramids of biomass are important examples of such methods.

If undisturbed by man, ecosystems develop over time until they reach a state of relative stability in which their diversity and adaptability is capable of adjusting to changes and to temporary disturbance such as fire, flood and disease. The influence of man, however, can shift this delicate balance. Our study of ecology must therefore include consideration of man's place in the control and management of ecosystems. The sources, effects and methods of limiting the damage caused by pollution are the obvious starting points, but we must also consider how to monitor and measure the effects of our activities. Awareness of the evidence offered by indicator species is vital, as is a thorough knowledge of the mistakes of the past in terms of bad management of ecosystems. Finally, we consider the control of largely artificial ecosystems such as those of agriculture and forestry.

FactZONE

It would be helpful to know the meaning of some biosphere words and phrases for the exam.

Use the next few pages to find the meanings of these words and then write them beside the correct definition. Tick them off as you go.

☐ habitat   ☐ producer   ☐ nitrate

☐ population   ☐ consumer   ☐ domestic pollution

☐ community   ☐ pyramid of biomass   ☐ micro-organism

☐ ecosystem   ☐ energy loss   ☐ organic waste

☐ abiotic factor   ☐ competition   ☐ indicator species

Words	Meanings
	a nitrogen-containing compound used by plants to manufacture protein
	a community and all the abiotic factors
	an organism that shows the level of pollution by its presence or absence
	energy used for movement and heat production
	the place where an animal or plant lives
	waste produced by homes and personal transport
	an organism that makes its own food by photosynthesis
	the dry weight of all the organisms at each stage in a food chain
	microscopic organisms such as bacteria and fungi
	all the members of one species in an area
	the result of two individuals needing the same scarce resource
	non-living conditions such as pH, oxygen concentration, temperature and light intensity
	an organism that eats another organism to gain energy
	unwanted material that was once part of a plant or animal
	all the animals and plants living in an area

Investigating an ecosystem

The main parts of an ecosystem are the animals, the plants and their habitat. When biologists investigate an ecosystem they need to know as much as possible about each of these parts.

Sampling for organisms

There are usually far too many organisms in an ecosystem to count them all. Biologists usually take a sample and assume that this represents the whole population.

Quadrats

A quadrat is used to sample organisms that do not move — usually plants. It is a frame that marks off an exact area so that plants in that area can be identified and counted.

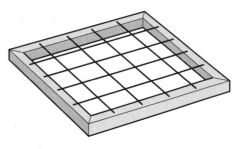

Quadrats must be placed randomly so that a representative sample is taken, and you should always look at several quadrats in an area to reduce the effect of unusual results. Your results are more reliable when you look at the results from many quadrats.

(tv) *Watch the video sequence showing how to sample with a quadrat.*

Pitfall traps

A pitfall trap can be used to sample small invertebrates living on the ground. You are likely to trap beetles and other insects as well as spiders and slugs.

Stones

> **REMEMBER** You need to be able to describe a technique used for sampling organisms.

Pitfall traps must be properly set up. They need to have their top flush with the ground surface so that the invertebrates are unaware of the trap before it is too late. As you can see in the diagram, it should be covered with a stone to keep out the rain, make it dark and stop birds eating your catch.

Traps should be checked often to avoid the animals escaping or being eaten before being counted. As with most methods, a large number of samples makes the results more reliable as it minimises the effects of unusual results.

Abiotic factors

Abiotic factors are non-living, physical conditions that can influence which plants or animals can live in an area. Some examples of abiotic factors include temperature, light intensity, water content of soil, the oxygen concentration in water and the level of pollutants.

 Can you think of any other abiotic factors?

REMEMBER You may be asked to identify two abiotic factors.

Measuring abiotic factors

Nowadays, there are scientific instruments available that allow us to measure most abiotic factors quickly and easily.

Light meters are used to measure light intensity. The meter is held at the soil surface and the sensor is pointed in the direction of maximum light intensity. Then the meter is read. As with any scientific measurements, we have to be aware of possible sources of error. The most common error is accidentally shading the light meter, or relying on too few readings.

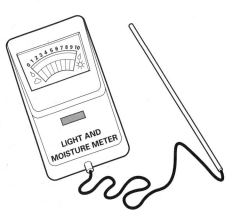

Soil moisture and pH meters are also available. Both are used by simply pushing the probe into the soil and reading the meter. This time the source of error might involve not cleaning the probe between readings. Damp soil clinging to the probe may give a false reading for the next sample.

Abiotic factors influence which plants and animals are found in a particular place. For example, some types of plants need shady places whilst others need full sunlight. Stonefly larvae can only survive in rivers with high concentrations of oxygen, whilst rat-tailed maggots can live in water with virtually no oxygen.

REMEMBER For the exam, you should be able to describe a technique used to measure an abiotic factor.

 Using the figure (right), say which abiotic factor may have been responsible for different plants growing in the field and beside the hedge.

In your exam you may be asked to explain how a particular abiotic factor influences the distribution of an organism. For example, grasses are found in full sunlight rather than in shady woodland because they need a high light intensity for photosynthesis.

How it works

First of all, it is important to be familiar with the names of important parts of a living system. Use the FactZone to be sure you know the following important terms: *habitat, population, community, ecosystem, producer, consumer.*

📺 Food chains and food webs

A *food chain* is a diagram that shows simple feeding relationships:

grass → rabbit → fox

is an example of a food chain.

❓ *Can you think of another food chain?*

A *food web* is a more complicated diagram like the one shown here – a simplified food web for a Scottish moorland..

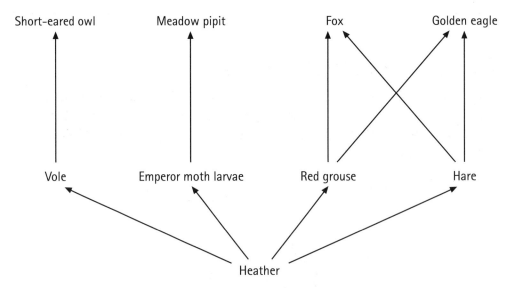

The arrows in a food web always show the *direction of energy flow*. If one species is lost from a food web, all the others will be affected — not just the predators or prey of that species. Energy is lost from a food web when it is used for *movement* or converted to *heat*.

Pyramids of numbers and biomass

A pyramid of numbers is simply a type of diagram that shows the number of organisms in each level of a food chain or web by a bar whose width is in scale with the number of organisms. It is really just a bar chart laid on its side.

The idea is that there should be lots of producers, then fewer primary consumers, then even fewer secondary consumers, and so on, which gives a diagram with a broad base and a narrow top — like a pyramid. The reason for the pyramid shape is that most of the food used in an ecosystem is used by the lower levels on the food web. The further up you go, the fewer animals there are, and the less food they eat. This is obvious really, because the food is made by the producers — green plants — and each level uses up most of the energy for its own movement and heat, so that only a small proportion can be passed on up the chain.

In practice, ecologists don't usually worry about *how many* organisms there are at each level. Instead they simply work out *the total weight* of all the producers, then the total weight of all the primary consumers, and so on. They don't include the weight of water in them because this varies a lot. We call the dry weight of anything its *biomass*. A *pyramid of biomass* is a diagram showing the total dry weight of the organisms at each level of a food web — and it is always a pyramid shape. The pyramid of biomass below is for a Scottish moorland.

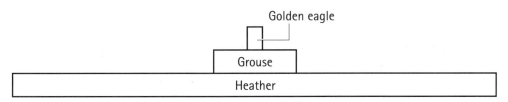

Population growth curves

The growth rate of any population depends on the birth rate and the death rate. Under ideal conditions, a population would have growth with a slow start, then a very fast rate of increase, and finally, the growth would slow down then stop.

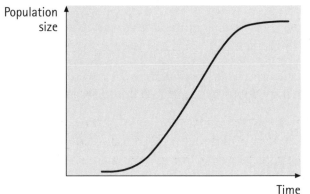

In a typical population growth curve:

- the first stage shows little growth because there are few organisms to reproduce

- a rapid increase follows as there is plentiful food and few predators

- the growth slows and stops as food supplies are used up and waste products build up in the habitat.

Factors that can limit the growth of a population include predation, shortage of food, disease, lack of water and lack of space.

Competition

Competition for resources that are in short supply is one of the most important factors in maintaining the stability and adaptability of ecosystems. For populations of any species to stay strong and stable over many generations it is essential that only the very best individuals reproduce and that the weak or less well-adapted are phased out. Competition is the main instrument by which this is achieved.

Competition in the same species

All the individuals in a species will have very similar needs. They will seek out the same type of food, habitat, shelter, nest sites and mates. Plants must compete for space, light, water and soil minerals. Any of these vital resources can be in limited supply and competition then becomes fierce. The best adapted individuals will often make sure that they have enough by defending *territories* or by establishing *dominance hierarchies*. The weaker individuals will be unable to get a share and will therefore migrate to other habitats. This makes sure that the population remains relatively stable at a sustainable level and that only the best individuals pass their genes on to the following generations.

Competition with other species

There are often times when different species make use of a single resource. For example, several different predators may take a particular prey species. If two species sharing the same habitat have needs that are too similar, then this will almost certainly lead to one of them being less able to compete and being driven from the habitat.

In fact, competition between species is usually quite limited in nature because species usually have different requirements.

 Make a list of pairs of species that might compete with each other for resources – for example, clover and grass might compete for light, water and nutrients.

The need for recycling

The chemical raw materials that are needed to make the tissues of living things are all recycled over and over again. If this did not happen, the world would have run out of these chemicals long ago. Decomposer organisms, mainly bacteria and fungi, release nutrients from dead bodies and waste material of animals and plants into the soil. These nutrients are then used by plants, which recycle them in the food web.

The nitrogen cycle is a particular example of this recycling, and you need to know a few details about how it works.

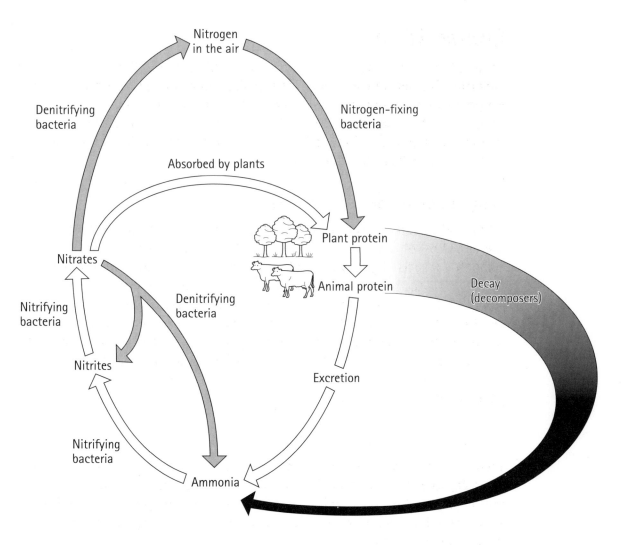

Nitrogen is a chemical element that is needed to make protein. The nitrogen in and around the Earth is constantly being recycled. It can be in the atmosphere, combined into chemicals in the soil, or be part of proteins in living things. The nitrogen cycle is a description of how it is changed from one form into other forms over and over again.

The protein in dead animals and plants (or their waste material) is converted into **ammonia** by bacteria and fungi, which eat the protein and produce ammonia as their waste product. Special bacteria then convert ammonia into **nitrites**. Other bacteria convert nitrites into **nitrates**. Plant roots absorb nitrates (they can't use ammonia or nitrites). The plants use the nitrates to make their protein and animals eat some of the plants to make their protein. Thanks to the micro-organisms in the soil, the cycle is completed and the nitrogen from waste and dead organisms has become part of many new living plants and animals.

This basic cycle happens continuously in the soil. In addition, there are some other factors that affect the recycling of nitrates and proteins:

- some bacteria (denitrifying bacteria) use up soil nitrates in such a way that the product escapes into the air as nitrogen gas
- some plants (such as peas, beans and clover) have nitrogen-fixing bacteria in their roots, which take nitrogen gas from the air and make it into nitrates.

Control and management

Sources of pollution

Pollution affects every ecosystem on Earth. Pollutants are found in the air, the land, fresh water and the sea. The three main sources of pollution are *industrial, agricultural* and *domestic*. You need to know an example of each. Industry releases sulphur dioxide, oil, heavy metals and radiation into the environment. Agricultural pollutants include pesticides, excess fertilisers and slurry. Our homes and cars produce sewage, detergents, sulphur dioxide and household waste.

Effects of energy sources on the environment

The energy needed by our modern society is steadily increasing. The two main ways in which electricity is generated are by burning fossil fuels and from nuclear power. There are drawbacks associated with both.

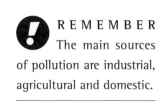 **REMEMBER**
The main sources of pollution are industrial, agricultural and domestic.

Burning fossil fuels creates soot and smogs that coat the environment with grime and restrict photosynthesis in plants. In addition, carbon dioxide and sulphur dioxide are released, contributing to global warming and acid rain. We are using up a resource that can never be replaced.

Nuclear power carries a risk of dangerous leaks of radiation at high levels after accidents. There is also a constant release of low level radiation in the cooling water from power stations. The used fuel has to be stored for thousands of years, and the long-term effects of storing radioactive materials are unknown.

Control of pollution

You need to know an example of how pollution may be controlled such as:

- world agreements to reduce carbon dioxide and sulphur dioxide emissions
- fitting catalytic converters to car exhausts
- treatment of sewage before releasing it into rivers.

◎ *How many materials can you think of that can be recycled?*

Pollution of water by organic waste

The graph at the top of page 17 shows the effect of a sewage discharge into a river. As you can see in the graph, unpolluted river water has few bacteria and a high oxygen content. A discharge of organic waste, such as sewage, changes all that. The bacteria use the sewage as food, multiply rapidly and use up all the oxygen.

Fish and most of the invertebrates in the river need a high oxygen concentration to survive.

The decrease in oxygen level results in many animals and plants dying or moving to an unpolluted area. Thus, sewage-polluted water supports a much poorer and less diverse community.

The graph shows that it is many kilometres downstream before all the sewage has been decomposed by the bacteria. By then, the bacteria numbers are low because of lack of food and the oxygen content has increased.

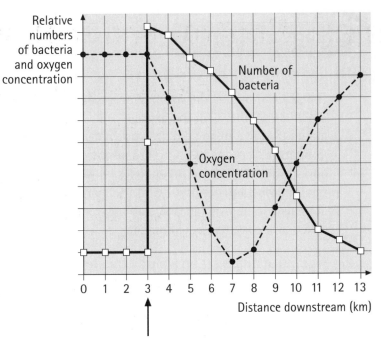

Indicator species

An indicator species is one whose presence or absence can be used to tell us something about the environment. For example, because the oxygen availability controls which species of animals can live in water, we can use the presence or absence of particular species to tell us how much pollution there is in a particular stretch of a river.

Results of poor management

You should know some examples of the damage that bad management can cause to ecosystems and how the situation might be improved.

- Stocks of certain fish species in the North Sea have almost disappeared because of overfishing. The stocks can be improved by setting limits on the numbers of fish that are allowed to be caught.

- The introduction of foreign species can have a disastrous effect on native ecosystems because of the lack of predators. For example, rabbits in Australia have turned huge areas of grassland into virtual desert.

Control of ecosystems by agriculture

When growing crops, a farmer is creating and controlling an artificial ecosystem. The soil must be kept fertile by using fertilisers. Pests and diseases are kept at bay with pesticides. Weeds compete with crop plants and so must be controlled with herbicides. However, many farmers value the native wildlife of the countryside and make efforts to manage their business in such a way as to allow habitats to remain for as wide a variety of species as possible.

Practice questions

1) Underline the correct option.

The place where an organism lives is called a *biosphere/habitat*. All the animals and plants that live in an area are known as a *population /community*. The organisms of one species that live in an area are known as a *population/community*. All the plants, animals and abiotic factors in an area are called an *ecosystem/environment*.

2) A group of pupils set up a pitfall trap to sample the invertebrates living on a woodland floor.

   a) Explain why the trap was covered with a stone.

   b) Suggest a possible reason to explain finding only beetles in the trap on the following day.

   c) How could this investigation be improved to obtain a more representative sample of the invertebrates living on the woodland floor?

3) Pupils used a quadrat in an investigation to estimate the abundance of limpets on a rocky shore. They counted 5 in their first quadrat. Then 7, 10, 3, 0, 3, 12, 5, 6 and 9 in the rest of the samples. Calculate the average number of limpets per quadrat.

4) Complete the following sentences.

Meters are used to measure abiotic factors. A .... meter is used to measure light .... The meter must not be .... by the body as this would give a reduced reading. When using a moisture meter, experimental error is reduced by .... the probe between samples. The .... of the results from both meters can be increased by taking several readings.

5) The following key may be used to identify different species of salmon.

i) no marks on side	pink
marks on side	go to ii)
ii) marks above lateral line	go to iii)
marks both sides of lateral line	go to iv)
iii) narrow marks	chum
round marks	sockeye
iv) narrow marks	silver
round marks	king

   a) Use the key to identify this salmon.

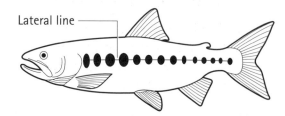

Lateral line

   b) Which characteristic could be used to distinguish a silver salmon from a chum?

6) The diagram below shows part of a food web from an oak wood.

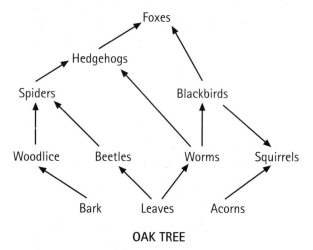

OAK TREE

   a) The oak tree is the producer in this food web. What is meant by the term producer?

   b) Select an example of a complete food chain consisting of five organisms from the food web above.

   .... → .... → .... →
   .... → ....

18

6) c) In terms of energy, what do the arrows in the food chain show?

d) Omnivores eat both plants and animals. Name the omnivore in this food web.

e) Unfortunately, the entire hedgehog family was killed in a local bonfire. Predict and explain the effect on the population of
i) spiders   ii) blackbirds

f) Explain how the worm and beetle are in competition with each other.

g) Not all the energy in a food web becomes available as energy for the next level. State two ways in which energy is lost from a food web.

7) Underline the correct option.

A pyramid of *energy/biomass* shows the *weight/numbers* of organisms in each stage of a food chain.

8) a) The recycling of nitrogen depends on a series of decomposer micro-organisms converting one substance to another.

The steps in a nitrogen cycle are listed below.
i)   pea roots absorb water and minerals
ii) dead pea leaves fall
iii) nitrites are produced
iv) pea plants make protein
v)   ammonium compounds are produced by decay of pea leaves.

Write the step numbers in the boxes below to complete a nitrogen cycle.

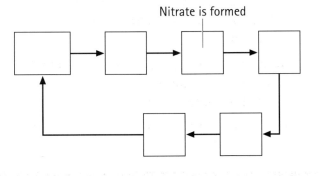

Nitrate is formed

b) The pea obtains nitrogen compounds from the activities of the decomposers. What advantage do the micro-organisms obtain from the cycle?

9) Pollution affects air, land and water.

a) Give one example of a pollutant from the following sources:
i)   agriculture
ii) industry
iii) domestic.

b) Describe an adverse effect of using fossil fuels.

10) The graph below shows the effect of an organic waste discharge into a river.

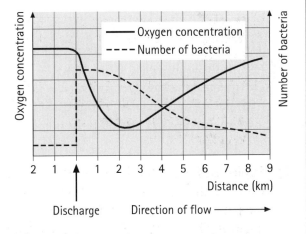

a) Give an example of an organic waste.

b) Account for the rise in the bacteria population.

c) Describe the relationship between the number of bacteria and the oxygen concentration.

d) Suggest where the smallest number of vertebrate and invertebrate species would be found.

11) Explain the meaning of the term indicator organisms. (Only the briefest definition is needed.)

The world of plants

This topic is about:

- **introducing plants**
  the variety of plants, uses of plants, consequences of loss of diversity

- **growing plants**
  seeds, flowers, pollination, fertilisation, seed dispersal, asexual reproduction

- **making food**
  transport systems, leaf structure, photosynthesis, limiting factors

There is an enormous range of plants on Earth, ranging from giant redwood trees over 100 metres tall to single-celled algae that can only be seen under a microscope. This variety is essential because it means that there is a large range of characteristics available for breeding from.

Plants are a vital part of life on Earth for all sorts of reasons. Animals need plants because they provide sources of food and offer a variety of shelter. In addition, by using up carbon dioxide and producing oxygen they help to maintain the balance of gases in the atmosphere.

From our point of view as humans, plants are endlessly useful. Examples are numerous, but it may help to bear in mind the following main categories of use:

- raw materials, e.g. timber for building

- foods, e.g. wheat for making bread

- medicines, e.g. foxgloves produce a medicine for treating heart problems.

◎ *Make a list of plants for each of the above categories. Look around and ask questions about the use of plants.*

We should also remember that plants form a delightful part of the beauty of our planet, both in natural ecosystems, in parks and gardens, and in our homes.

If any species of plant is allowed to die out, the possible consequences are serious both for man and for the other living things that share our planet. Many plants represent potential resources (food or raw materials) that may become essential in the future. Lots of plants may contain valuable products that have not been discovered yet. Every plant provides food and shelter for a variety of other organisms, some of which may only be able to live on that particular species. The genetic characteristics of a particular plant may turn out to be very useful. Biologists can transfer such characteristics from one species to another, and the loss of any species means a reduction in the total reserves of available genes.

It would be helpful to know the meaning of some plant words and phrases for the exam.

Use the next few pages to find the meanings of the words and then write them beside the correct definition. Tick them off as you go.

- ☐ stoma
- ☐ fertilisation
- ☐ clone
- ☐ pollination
- ☐ raw materials
- ☐ asexual
- ☐ stigma
- ☐ chlorophyll
- ☐ stamen
- ☐ xylem
- ☐ refining process
- ☐ phloem
- ☐ palisade mesophyll
- ☐ structural carbohydrate
- ☐ limiting factor

Words	Meanings
	joining of the nuclei from a pollen grain and an ovule
	part of a flower where the pollen lands
	living cells that form a tube to carry food from the leaves to other parts of the plant
	a manufacturing process that converts part of a plant into a saleable product such as jam or planks
	the type of reproduction in which one parent produces genetically identical offspring
	a pore in the leaf epidermis that allows carbon dioxide to diffuse in for photosynthesis, and oxygen and water vapour to diffuse out
	the leaf cells in which the majority of photosynthesis occurs
	makes and releases pollen
	the molecule in chloroplasts that traps light energy
	a group of genetically identical plants.
	the transfer of pollen from the stamen to the stigma by wind or insects
	vessels that carry water from the roots to the leaves
	parts of plants used in a manufacturing process
	something that slows the speed of photosynthesis because it is in short supply
	cellulose that is used by plants to make cell walls

Growing plants

📺 Seeds

The seed of a dicotyledonous plant has three main parts:

- the *seed coat*, which is a tough protective outer covering

- the *embryo*, consisting of the young root and the young shoot, which will develop into the adult plant

- the *food store* which provides energy and raw materials for the young plant to use until it is large enough to make its own food.

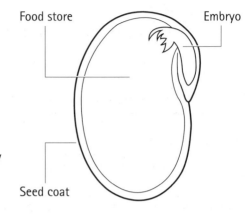

Food store   Embryo

Seed coat

Germination is the first stage of the development of a new plant from the seed. Three factors are required for successful germination:

- *water* – which allows the seed to swell up and the embryo to start growing

- *oxygen* – to allow energy to be released during germination

- *a suitable temperature* – almost no seeds will germinate at temperatures below 5°C. High temperatures (above 45°C) also prevent germination. Between 5°C and 45°C germination percentage increases with temperature up to a point where germination is at its best – the optimum temperature.

Flowers

Flowers are the organs of sexual reproduction in plants. You need to be familiar with the parts of the flower shown in the diagram, and what function each part has.

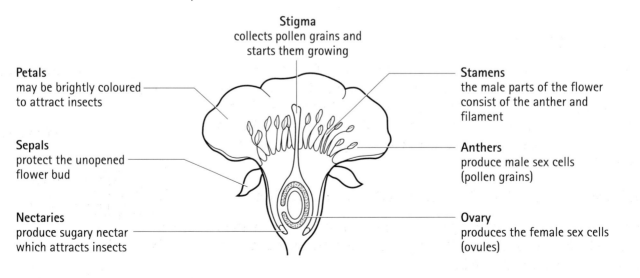

Stigma
collects pollen grains and
starts them growing

Petals
may be brightly coloured
to attract insects

Sepals
protect the unopened
flower bud

Nectaries
produce sugary nectar
which attracts insects

Stamens
the male parts of the flower
consist of the anther and
filament

Anthers
produce male sex cells
(pollen grains)

Ovary
produces the female sex cells
(ovules)

⊡ Pollination

Pollination is the transfer of pollen from the anthers of one flower to the stigmas of another. *Wind* and *insects* are the main pollinators. Wind-pollinated flowers are very different in structure from insect-pollinated flowers.

Insect-pollinated	Wind-pollinated
large, brightly coloured petals – to attract insects	small petals, often brown or dull green – no need to attract insects
often sweetly-scented – to attract insects	no scent – no need to attract insects
usually contain nectar – to attract insects	no nectar – no need to attract insects
moderate quantity of pollen – less wastage than with wind pollination	pollen produced in great quantities – because most does not reach another flower
pollen often sticky or spiky – to stick to insects	pollen very light and smooth – so it can be blown in the wind
anthers firm and inside flower – to brush insects	anthers loosely attached and dangle out – to catch the wind
stigma inside the flower – so that the insect brushes against it	stigma hangs outside the flower – to catch the drifting pollen
stigma has sticky coating – pollen sticks to it	stigma feathery or net-like– to catch the drifting pollen
Insect-pollinated	Wind-pollinated

Fertilisation

When pollen grains land on the stigma of a flower of the correct species they germinate. A pollen tube grows through the tissues of the flower until it reaches an ovule inside the ovary. The nucleus of the pollen grain (the male gamete) then passes along the pollen tube and joins with the nucleus of the ovule (the female gamete). This process is called *fertilisation*.

⊙ *Refer to the website for an animated diagram of fertilisation.*

After fertilisation, the female parts of the flower develop into a fruit. The ovules become seeds, and the ovary wall becomes the rest of the fruit that surrounds the seeds.

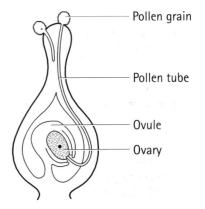

Pollen grain

Pollen tube

Ovule

Ovary

📺 Seed dispersal

It is vital that seeds are dispersed away from each other and from the parent plant so that there is less competition for light, water and nutrients. The commonest methods of seed dispersal are:

- *wind*, e.g. dandelion, sycamore, thistle, willow herb — fruits are light and have extensions that act as parachutes or wings to catch the wind

- *animal – internal*, e.g. tomato, plum, raspberry, grape — have brightly-coloured and succulent fruits that contain seeds with indigestible coats, which allow the seeds to pass through the animal undamaged

- *animal – external*, e.g. goose grass, burdock — the fruits have hooks that attach them to the fur of passing animals

? *There are two of each type in the diagram below. Can you say which type of dispersal is used by the fruits in each picture?*

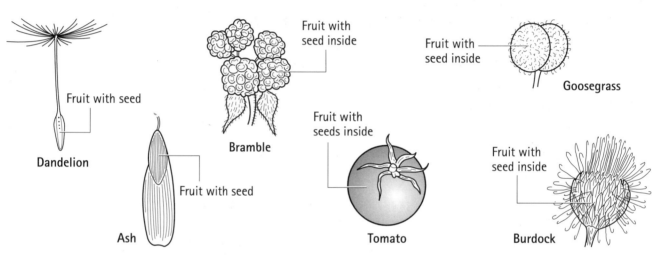

Fruit with seed — Dandelion

Fruit with seed — Ash

Fruit with seed inside — Bramble

Fruit with seeds inside — Tomato

Fruit with seed inside — Goosegrass

Fruit with seed inside — Burdock

Asexual reproduction

Plants can also reproduce asexually (i.e. without flowers or fertilisation). There are many ways in which this is achieved. Some are natural, whereas others have been developed by people. Plants that are produced in these ways are genetically identical to each other and to the parent; a group like this is called a *clone*.

Natural methods

The most important natural methods of asexual reproduction are:

- *runners* — these are side shoots that grow out from the parent plant and form buds at points along the runner; the buds form roots and grow into new plants, e.g. strawberry, spider plant

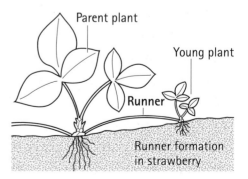

Parent plant

Young plant

Runner

Runner formation in strawberry

- *tubers* – a tuber is an underground food store that stores enough food over winter to start the growth of a new plant; each new plant can then make its own food and form several tubers for the next winter, e.g. potatoes, dahlias

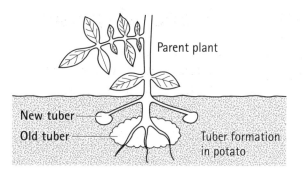

Parent plant

New tuber

Old tuber

Tuber formation in potato

(?) *Look at, and think about, all the plants you eat. What part of the plant are they from? Are they a reproductive structure? If so, how would they be dispersed, or would they reproduce asexually?*

Artificial methods

Humans are able to propagate plants artificially. *Taking cuttings* and *grafting* are the methods of artificial propagation that you need to know:

- *cuttings* – a piece of stem with some leaves attached is taken from the parent plant and placed in suitable compost where it can form roots

- *grafting* – a bud or shoot of a valuable plant is taken and joined to a plant that already has a root system

Artificial methods of propagating plants have many advantages to man:

- they are a quick method of propagating large numbers of plants

- uniformity is guaranteed

- once a new variety is produced (by sexual reproduction), a large supply of identical plants can be produced for selling

- rare plants can sometimes be saved from extinction by artificial propagation.

Sex or no sex?

Both sexual and asexual reproduction have advantages for plants – each method is used when it best suits the plant.

Advantages of asexual reproduction:

- only one parent plant is required

- young plants get food and water from the parent and so can grow quicker

- young plants are bigger and less vulnerable than tiny seedlings

- young plants are identical to the parent – good features are passed on.

Advantages of sexual reproduction:

- characteristics are inherited from two parents and this produces variation, e.g. less chance of disease, ability to adapt to changes of climate etc.

- seeds are tough and can stay dormant until conditions are favourable

- the offspring can be dispersed further from the parent plant – and possibly colonise new habitats.

Making food

Transport systems

Plants need both water (from the roots) and dissolved food (manufactured in the leaves) in all parts of the plant. Therefore, they need transport systems to move these substances around the plant. *Xylem* and *phloem* are the two main transport systems.

- Xylem carries water, which is drawn in from the soil by root hairs, upwards to all parts of the plant.
- Phloem carries dissolved food (sugars) from the leaves in all directions to parts of the plant that cannot make their own food (growing tips, roots, storage organs, flowers, developing fruits, etc.).

Xylem facts

Xylem vessels are dead. The end walls have disintegrated to leave hollow tubes, and the side walls have become impregnated with rings of lignin, which strengthen the tubes. Xylem also carries dissolved minerals and helps to support the plant.

◎ *Look at the wooden furniture in your room. Most of what you can see is xylem.*

Phloem facts

All phloem cells are alive. Separate companion cells provide the energy for the tube cells. The end walls of the tube cells have pores. Food is transported through the pores from cell to cell in the form of dissolved sugars.

Leaf structure

The leaf is the organ that plants use to make food by photosynthesis. You must be familiar with the structure of the leaf and the function of the parts shown in the diagram on page 27.

Stomata are tiny pores on the surface of leaves that allow gases in and out of the leaf. Plants take in carbon dioxide gas from the air through the stomata and get rid of the excess oxygen produced by photosynthesis. Because a lot of water vapour can be lost through the stomata, they only open for photosynthesis in daylight; at night, they close to reduce loss of water vapour.

◎ *Leaves come in many shapes and sizes. Look closely at different leaves and decide whether some leaf shapes and textures might be better at reducing water loss compared with others.*

Veins
contain xylem (top part of vein) for water transport and phloem (lower part) to transport dissolved food

The epidermis
protects the leaf and is transparent to let light through

Palisade mesophyll
cells are tall and closely packed to absorb maximum light cells contain many chloroplasts most photosynthesis takes place in the palisade cells

Spongy mesophyll
also captures light and makes food cells have air spaces between them to allow easy gas exchange

Air spaces
allow gases to get into and out of every cell in the leaf

Stomata
(mostly on the underside of the leaf) allow gas exchange

Photosynthesis

Green leaves use light energy to combine carbon dioxide and water together to make glucose and oxygen. This process is called photosynthesis. Chlorophyll is a green chemical found in plant cells and is essential for photosynthesis. Chlorophyll captures light energy from the sun and converts it into chemical energy, which is used to combine carbon dioxide with water to form glucose.

Photosynthesis is summarised in the word equation below.

carbon dioxide + water → glucose + oxygen

Storage and structural carbohydrates

The food that plants make is in the form of *carbohydrates*. Carbohydrates are chemicals containing only the elements carbon, hydrogen and oxygen

The simplest useful form of carbohydrate produced by photosynthesis is the simple sugar called glucose. Glucose may be used as an energy source, or it may be converted into other carbohydrates such as *starch* for storage, *cellulose*, which forms plant cell walls, or *lignin*. Cellulose and lignin form part of the structure of the plant and are therefore sometimes referred to as structural carbohydrates.

Limiting factors

A limiting factor is a factor that slows down or stops a process because it is in short supply. Photosynthesis can often be limited because certain factors are in short supply. The commonest limiting factors in photosynthesis are *low light intensity*, *lack of carbon dioxide* and *low temperature*.

Practice questions

1) Because of the activities of man, such as deforestation, the number of plant species in the world is decreasing. What are the consequences for

    a) man   b) other animals?

2) Decide which of the following describes the use of plants for food, medicines or raw materials.

Example A
Bamboo plants commonly grow wild in the Far East. Gathered cheaply, the canes are put to many uses that in the West would demand steel or other man-made materials. Houses, scaffolding, water pipes and furniture are all constructed with this renewable resource.

Example B
A little-known grain plant once grown by the Aztecs is awakening renewed interest. Amaranth has high protein levels. In addition, it contains lysine, an essential amino acid which is lacking in our staple cereal crops.

Example C
The native Indians of North America used the purple cone flower, *Echinacea*, as a cure for many maladies. It has been rediscovered as a remedy for acne and other skin disorders.

3) Describe how barley is refined in the malting process.

4) Give an example of a potential use of plants or plant products.

5) a) Describe the function of the following seed parts:
      i) seed coat
      ii) embryo
      iii) food store

    b) What are the three conditions essential for the germination of seeds?

6) The diagram below shows the structure of a grass flower.

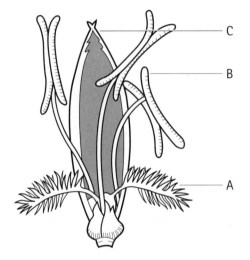

    a) Which letters label the i) petal, ii) stigma and iii) anther?

    b) Describe how wind pollination is made more efficient by the shape and position of the:
      i) stigma
      ii) anthers.

    c) What is the meaning of the term 'pollination'?

    d) Most wind-pollinated flowers have green petals. Why are insect-pollinated flowers any colour except green?

7) Complete the following sentences to describe fertilisation.

Soon after the pollen grain lands on the .... it germinates and begins to grow a pollen .... This grows down into the ovule inside the .... The pollen .... then flows down to fuse with the ovule .... This fusion is known as fertilisation.

8) Decide which of the following mechanisms describes wind, animal-internal or animal-external fruit dispersal.

Example A
Brambles are attractive to many birds when ripe and black. A day or so later, the seeds are deposited miles away in the birds' droppings.

Example B
Rosebay willowherb produces tiny fruits with long white 'hairs'. It is also called 'fireweed' because it quickly colonises freshly-burned ground.

Example C
Goosegrass has tiny hooks on its fruits, leaves and stem. These cling to clothes and fur, and the carrier may have gone miles before they are brushed off.

9) Plants may reproduce asexually or sexually.
   a) Give an example of a plant that reproduces asexually.
   b) Describe one advantage of asexual reproduction in plants.
   c) What is the meaning of the term 'clone'?
   d) Describe one advantage of sexual reproduction to plants.

10) a) The answer to the following questions is either xylem or phloem.
   Which type of vessel:
   i) has walls impregnated with waterproof lignin?
   ii) carries dissolved food from leaves to other parts of the plant?
   iii) has sieve plates?
   iv) is composed of living cells?
   v) is a hollow tube?
   b) Apart from transport, describe another function of xylem.

11) The diagram below shows the microscopic structure of a leaf.

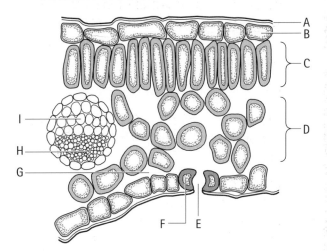

   a) Name the parts labelled B, E and I.
   b) Which letters label:
      i) guard cells?
      ii) cells where most photosynthesis occurs?
   c) What is the function of part A?
   d) Name the gas, needed for photosynthesis, that diffuses in through E.
   e) Which two gases diffuse out through E?

12) Complete the following word equation.

   ? + water $\xrightarrow[\text{chloropyll}]{\text{light}}$ food + ?

13) Complete the following sentences.
   Chlorophyll changes .... energy to .... energy. Using this energy, carbon dioxide is converted to the storage carbohydrate known as .... and the structural carbohydrate called ....

Animal survival

This topic is about:

- **the need for food**
  types of food, digestion, teeth, the digestive system, peristalsis, digestive enzymes, absorption

- **reproduction**
  sperm and eggs, internal and external fertilisation, reproductive organs, development in mammals, parental care

- **water and waste**
  water balance, structure of the kidneys, kidney failure

- **responding to the environment**
  response to environmental factors, rhythmical behaviours

All living organisms have particular requirements to ensure survival and reproduction. In this topic, we will consider four of the most important.

Food provides the raw materials needed for growth and development, and the energy required to sustain all activity. In order to make the variety of foods that are eaten available in a form that cells can use, food must be digested, dissolved and absorbed. The structure of the **digestive system**, and the way in which it works, is therefore a prime essential for animal survival.

Reproduction is the means by which organisms replace dead individuals. It provides increased numbers for potential colonisation and spreading into new habitats. In addition, by the constant process of death and renewal, essential variation is produced. This allows for the selection of the best and fittest from each generation to ensure the best chances of survival and evolution. In particular, we will focus on the processes of mating, fertilisation in fish and mammals, the reproductive organs of mammals, protection of the embryo and care of the young.

Water is critical to all life forms, and the maintenance of an internal water balance is a particular requirement. We will consider the structure and function of the **kidney** and its ability to adjust to the varying rates of water intake and output. In addition, we will look at its role in the excretion of waste materials. The importance of the kidneys to human health, and the benefits and drawbacks of treatments, such as dialysis and transplantation, will also be considered.

The final aspect of animal survival that we will examine is **behaviour**. It becomes completely irrelevant how efficient and fit your body is if you have a tendency to upset tigers, or to walk under buses. Behaviour has a very direct influence on survival! The ways in which stimuli become associated with appropriate responses will be considered along with some aspects of rhythmical behaviours and the effect of the environment on behaviour.

FactZONE

It would be helpful to know the meaning of some animal survival words and phrases for the exam.

Use the next few pages to find the meanings of the words and then write them beside the correct definition. Tick them off as you go.

- ☐ protein
- ☐ digestion
- ☐ lipase
- ☐ peristalsis
- ☐ large intestine
- ☐ external fertilisation
- ☐ testes
- ☐ yolk sac
- ☐ amniotic sac
- ☐ metabolic water
- ☐ renal artery
- ☐ glomerulus
- ☐ urea
- ☐ chemical stimulus
- ☐ rhythmical behaviour

Words	Meanings
	contraction of muscles behind food, which pushes it along the gut
	the vessel that carries blood to the kidneys
	a molecule that contains carbon, hydrogen, oxygen and nitrogen, and is composed of many amino acids joined together
	make sperm
	food supply for a newly-hatched fish
	the enzyme that digests fat into fatty acids and glycerol
	made from excess amino acids, which forms the main waste substance in urine
	mating or migration or hibernation
	breakdown of large, insoluble molecules to small, soluble molecules
	the environmental factor that attracts *Planarian* worms to rotting meat
	a bag of water that protects a fetus
	part of the digestive system that absorbs water
	sperms and eggs joining outside the body
	a knot of capillaries at the beginning of a nephron
	a byproduct of aerobic respiration

The need for food

All organisms need food for growth and energy.

Types of food

Our food is made up from three main components. These are *carbohydrates*, *fats* and *proteins*.

Carbohydrates are built up from simple sugar units. They may be single units such as glucose, or long chains of sugar units joined together, such as in starch. Each fat molecule is made from glycerol combined with three fatty acids. Proteins are made up of long chains of amino acids linked together.

The table below lists the facts you need to know about the food components.

Food component	Elements present	Sub-units	Sub-units linked together	Function in the body
carbohydrate	carbon, hydrogen and oxygen	simple sugars		source of energy
fat	carbon, hydrogen and oxygen	glycerol and fatty acids		source of energy
protein	carbon, hydrogen, oxygen PLUS nitrogen	amino acids		for growth and for repair of damaged or worn-out tissues

◎ *Complete the following sentences.*

Carbohydrates and fats contain the elements ...., ... and .... In addition protein contains ....

Carbohydrates are made up of .... joined together, proteins are long chains of .... and fats are built from .... and .... linked together.

Digestion

Digestion is the process of breaking down large food particles into particles that are small enough to diffuse through the wall of the small intestine into the blood stream.

Teeth

Digestion begins in the mouth. Teeth are used to bite off manageable lumps of food. They then crush the food pieces and mix them with saliva.

The table below shows how teeth are adapted for different diets.

🔮 *Check your understanding with the Test bite on the Bitesize website.*

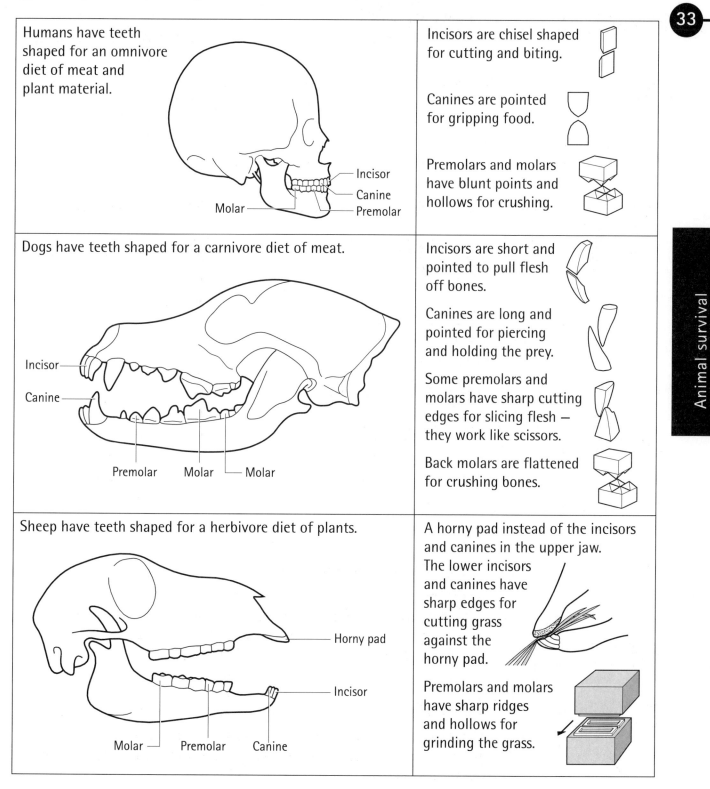

Humans have teeth shaped for an omnivore diet of meat and plant material.	Incisors are chisel shaped for cutting and biting.
	Canines are pointed for gripping food.
Molar · Incisor · Canine · Premolar	Premolars and molars have blunt points and hollows for crushing.
Dogs have teeth shaped for a carnivore diet of meat.	Incisors are short and pointed to pull flesh off bones.
Incisor · Canine · Premolar · Molar · Molar	Canines are long and pointed for piercing and holding the prey.
	Some premolars and molars have sharp cutting edges for slicing flesh — they work like scissors.
	Back molars are flattened for crushing bones.
Sheep have teeth shaped for a herbivore diet of plants.	A horny pad instead of the incisors and canines in the upper jaw. The lower incisors and canines have sharp edges for cutting grass against the horny pad.
Horny pad · Incisor · Molar · Premolar · Canine	Premolars and molars have sharp ridges and hollows for grinding the grass.

Animal survival

The digestive system

After mechanical breakdown by the teeth, food enters the digestive system for chemical breakdown. This is where a series of enzymes break apart the large food molecules. Protein, fat and starch in food are large insoluble molecules. Digestive enzymes break these down into small, soluble molecules.

The diagram below shows the parts of the alimentary canal you may be asked to identify and name in the exam.

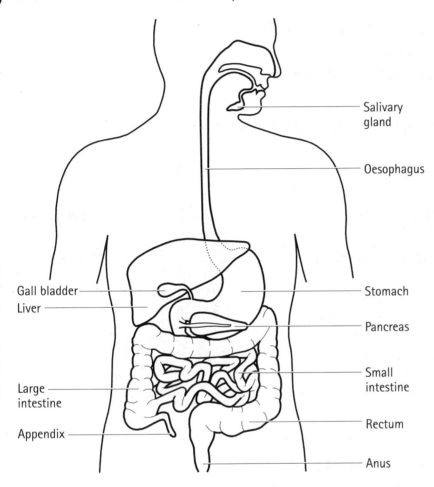

Salivary gland

Oesophagus

Gall bladder

Liver

Stomach

Pancreas

Large intestine

Small intestine

Appendix

Rectum

Anus

Digestive juices are needed for digestion to work quickly. These juices are produced by salivary glands, the stomach, liver, pancreas and small intestine.

◎ *Colour in the parts of the gut that produce digestive juices.*

Contractions of the muscular stomach wall help in the chemical breakdown of food. They mix the food thoroughly with pepsin and acid in the gastric juice, which speeds up the digestion of food.

Peristalsis

Peristalsis is the way in which food is pushed through the gut. Muscles in the wall of the gut, around and in front of the food, relax whilst the muscles behind the food contract, pushing it along.

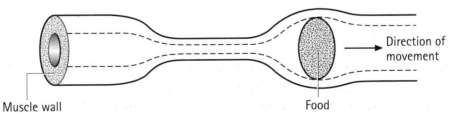

Direction of movement

Muscle wall

Food

📺 *Watch the video sequence about the digestive system and peristalsis.*

Digestive enzymes

Enzymes are chemicals that speed up biological reactions. Digestion involves several enzymes that speed up the breakdown of food. Different enzymes are needed for the breakdown of proteins, carbohydrates and fats.

The enzyme details you need to know are:

- amylase enzymes break down starch into simple sugars
- there is an amylase in saliva
- protease enzymes break down proteins into peptides and amino acids
- pepsin (made in the stomach) is a protease
- lipase enzymes break down fats into fatty acids and glycerol
- lipase is made in the pancreas.

◎ *Write the word equations for the action of amylase, pepsin and lipase.*

! **R E M E M B E R**
Only small soluble molecules, such as glucose, amino acids, fatty acids and glycerol, can diffuse out of the gut into the villi.

35

Absorption

Absorption is the process by which digested food diffuses from the digestive system into the blood stream.

(?) *In what form will the starch in a piece of bread be by the time it reaches the small intestine?*

Absorption happens mostly through the walls of the small intestine, which is very well adapted for the purpose. The small intestine is long and has a folded lining to create a large surface area. The surface area of the small intestine is further increased by microscopic finger-like projections called villi. The diagram (right) shows the parts you need to learn.

— Blood capillary

— Lacteal

— Wall

The features of the villi that make them efficient at absorption are:

- millions of villi provide a large surface area for absorption
- their walls are only one cell thick, which allows rapid diffusion
- they contain many blood capillaries that absorb glucose and amino acids
- they contain lacteals that absorb fatty acids and glycerol
- both the blood capillaries and lacteals are connected to a network of vessels that allow transport of food to the rest of the body.

The large intestine

The wall of the large intestine absorbs most of the water from the undigested food into the blood system. The undigested remains, called faeces, are stored in the rectum. When convenient, the faeces are passed out of the body through the anus. This is called elimination.

! **R E M E M B E R**
Air sacs and capillaries also have a large surface area to make them efficient for their function.

Animal survival

Sperm and eggs

Animals produce special cells for sexual reproduction. The male sex cell is called a sperm. It has a head, a nucleus and a tail so that it can swim. The female sex cell is called an egg. It contains a nucleus and a store of food. It is much larger than a sperm.

Sperm cell

Egg cell

Fertilisation

Fertilisation is the joining of the nucleus of a single sperm with the nucleus of an egg cell. The egg then forms a strong membrane to prevent other sperm from entering.

External fertilisation — fish

Fish produce a large number of sex cells, which are released into the water. Some species of fish have courtship rituals that make sure that the male and female sex cells are released near each other. This increases the chance of fertilisation and so fewer eggs are needed.

(?) *A trout produces 1000 eggs and a cod more than 500,000. Suggest which fish has the more effective courtship behaviour.*

Fish eggs consist of a soft protective covering containing the embryo and a food supply. The embryo develops inside the egg using only the energy stored in the yolk.

The picture shows a newly-hatched fish, which has a yolk sac to provide food for a few days until it starts to feed.

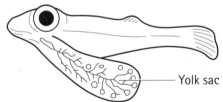

Yolk sac

 REMEMBER
Sperm need water to swim in. They can swim in the water outside a fish; but they need water inside the female of land-living animals.

Internal fertilisation — mammals

Animals that live on land have internal fertilisation. This is necessary because sperm need fluid in which to swim to the egg. Internal fertilisation is more efficient than external fertilisation. The chance of fertilisation is much greater because the sex cells are closer together and protected when released. This means that fewer eggs need to be produced.

Reproductive organs

The diagrams below show the male and female reproductive systems in humans. Sperm cells are produced in testes, travel down the sperm ducts and are introduced into the vagina by the penis. Egg cells are produced in ovaries and are then released into the oviducts. Fertilisation takes place in the oviducts. The fertilised egg cell then starts to divide, producing a ball of cells that moves down the oviduct and into the uterus, where it attaches itself to the lining of the uterine wall.

◎ Mark a cross on the diagram to show where fertilisation takes place.

◉ Watch the fertilisation movie.

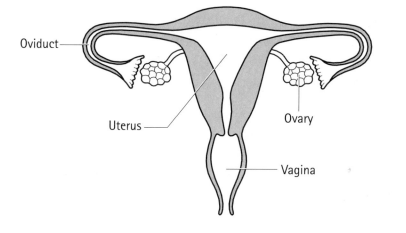

Oviduct

Uterus

Ovary

Vagina

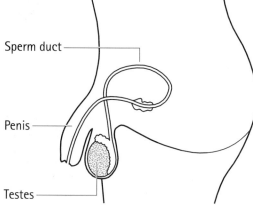

Sperm duct

Penis

Testes

Mammalian development

The picture shows a fetus developing inside the uterus. You may be asked to identify and name any of the labelled parts. The fetus grows in a bag of watery fluid called the amniotic sac. This provides protection from knocks. The fetus is connected to the placenta by the umbilical cord. The placenta allows food and oxygen to diffuse from the mother's blood system to the growing fetus. Waste materials and carbon dioxide diffuse in the other direction and are removed in the mother's blood.

Parental care

The number of eggs that any animal needs to produce depends on the chance of fertilisation and the degree of parental care. External fertilisation and the lack of parental care means that large numbers of eggs must be produced by most fish. Mammals provide protection during fertilisation and development, and then food and care after birth. The chance of a mammal egg being fertilised and then surviving to beyond birth is high. Thus mammals only need to produce a few eggs to ensure the survival of the species.

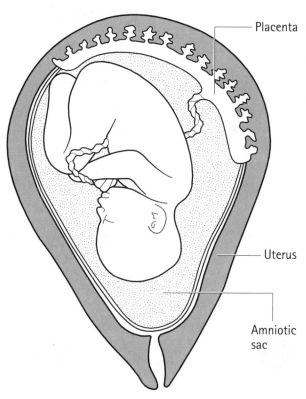

Placenta

Uterus

Amniotic sac

Animal survival

Water and waste

The water concentration of blood must be kept constant. Poisonous wastes must be removed from the body before they build up to harmful concentrations. Both these crucial tasks are done by the kidney.

Water balance

The water lost must be equalled by the water taken in. The table shows the ways in which mammals lose and gain water.

Water loss	Water gain
breathing	food
sweat	drink
faeces	water from respiration
urine	

Structure of the kidneys

The diagram shows the kidneys, bladder and renal blood vessels. You need to learn all the labelled parts.

Blood is brought to the kidney in the renal artery. The kidneys *filter* the blood and then *reabsorb* useful materials, such as glucose. After it has been purified, the blood returns to the circulation through the renal vein.

Urine is taken from the kidneys to the bladder by the ureter. The bladder stores the urine until it is convenient to expel it from the body.

 Watch the video sequence of a real kidney being dissected.

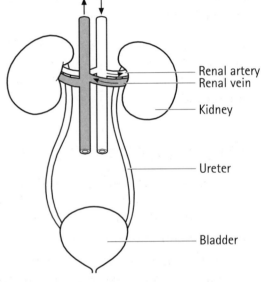

Blood flow

Renal artery
Renal vein

Kidney

Ureter

Bladder

Kidney failure

Kidneys may be damaged by accidents or disease. Kidney failure can lead to a build up of poisonous wastes in the body, and can be fatal. Treatment is by dialysis on a kidney machine, or by a kidney transplant.

Kidney machines can keep patients alive until a transplant is possible but have several disadvantages. They are expensive and the patient must have his or her blood connected to the machine for many hours every week.

After a transplant, the patient can lead a normal life. However, any major surgery carries some risk, drugs need to be taken for a long period to prevent the kidney being rejected, and there is a severe shortage of donors.

 R E M E M B E R In the exam, you may be asked to describe both the benefits and limitations of replacement or 'artificial' kidneys .

The nephron

Urine is produced in microscopic structures in the kidney called nephrons. Each kidney has many millions of nephrons.

The diagram shows a single kidney nephron. You should be able to identify and name all the labelled parts.

Blood is filtered by the glomerulus. Water and small dissolved molecules, such as glucose, salt and urea, are forced out of the glomerulus capillaries. Blood cells and large molecules, such as protein, are too big to be filtered out.

The filtrate is collected by the Bowman's capsule and enters the tubules. Useful substances, such as glucose, water and some salt, are reabsorbed into the blood. Reabsorption is through blood capillaries, which you can see are closely wrapped round the tubules.

The waste, consisting of water, some salt and urea, is called urine. The urine is collected by the collecting duct, taken to the ureters and then to the bladder.

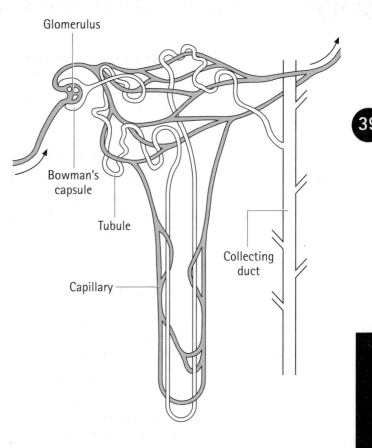

Glomerulus

Bowman's capsule

Tubule

Collecting duct

Capillary

Control of water balance

The kidneys are the main organs that a mammal uses to adjust its water content by controlling the volume of urine produced. The kidneys are controlled by ADH (anti-diuretic hormone). The brain detects any changes in the water concentration of the blood.

■ If the blood water concentration rises, then less ADH is released and so the kidney reabsorbs less water. This means that a large volume of very dilute urine is produced. This returns the blood water concentration to normal.

■ If the blood water concentration falls, then more ADH is released. In response to this the kidney reabsorbs more water. This means that a small volume of very concentrated urine is produced. This returns the blood water concentration to normal.

REMEMBER When we drink more than we need, less ADH is released and so we produce a greater volume of urine.

Urea — a poisonous waste

The main waste product removed in urine is urea. Urea is produced when surplus amino acids are broken down. Urea is produced in the liver and then carried away in the blood. It is filtered out and excreted by the kidneys.

REMEMBER When we sweat, more ADH is released and so we produce a smaller volume of urine.

Response to environmental factors

All animals must react to changes in their environment if they are to survive. **Light**, **humidity** and **chemicals** are three environmental factors that affect the behaviour of animals.

Light

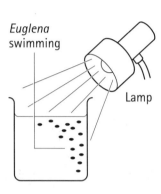

Euglena swimming

Lamp

The diagram shows an experiment into the effect of light on a one-celled organism called *Euglena*. As you can see, *Euglena* swims towards the light. This behaviour benefits *Euglena* because it contains chloroplasts and so can use the light for photosynthesis.

Other organisms move away from the light into a darker place. Woodlice move away from light and hide beneath stones or logs. By doing so, they become less visible to predators.

(?) *Maggots feed by burrowing through the flesh of dead animals. Are they likely to move towards or away from light?*

Humidity

The diagram below shows a choice chamber used to find out if woodlice spend more time in dry or damp conditions. They are found to spend the majority of their time in the humid side. By moving towards humid conditions, the woodlice will keep their gills moist and able to work efficiently.

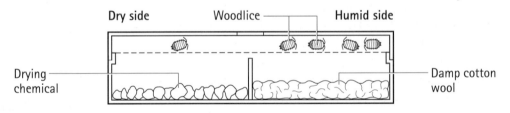

Dry side   Woodlice   **Humid side**

Drying chemical   Damp cotton wool

Chemicals

The diagram below shows an experiment to find the effect of liver on the behaviour of *Planarian* worms.

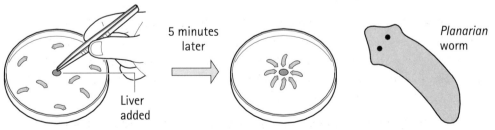

5 minutes later

Liver added

Planarian worm

The worm is attracted by the chemicals diffusing from the meat. This behaviour benefits the animal as it will move towards food material. By responding to smells, animals may find food, avoid danger, or find potential mates. In every case, the response is important for survival.

40

Watch the video sequence showing scallops responding to chemicals from a starfish.

Rhythmical behaviour

Rhythmical behaviour in animals is triggered by regular events in the environment, such as day and night, tides or seasons.

Daily rhythms

Cockroaches have a daily activity cycle controlled by darkness. They are active at night and hide during the day. This behaviour is to their advantage as they can feed much more safely under cover of darkness.

 Think of other examples of animals that are active only during the night.

Tidal rhythms

Activity in many seashore animals is triggered by the tides. Many animals can only feed when covered by the sea and remain inactive when the tide is out.

The picture shows a fiddler crab, which hides in a burrow when the tide is in. They come out at low tide to feed on what has been left behind by the sea.

Seasonal rhythms

Breeding, migration and hibernation are all triggered by day length.

The graph shows how the number of daylight hours affects the mating activity in sheep. The sheep start to breed in autumn as the number of daylight hours decreases. This means that the lambs will be born in the spring. This behaviour is to the advantage of the sheep because their lambs will be born at a time of year when the weather should be warmer and food plentiful.

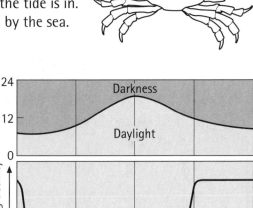

 Watch the video sequence that shows how the increasing day length of spring synchronises the growth of leaves, the hatching of caterpillars and the breeding of blue tits.

The migration of birds is triggered by changing day length. Swallows, for example, feed on insects that are available in Scotland only in the warmer months of the year. If they remained in Scotland during winter, they would certainly starve. In autumn, the shortening days trigger migration to Africa where there is a plentiful supply of insects.

Bats and hedgehogs would also be likely to starve in the Scottish winter. The shortening days in autumn trigger a different behaviour – these animals go into hibernation until spring.

Animal survival

Practice questions

1) Complete these sentences.

All food types contain carbon, hydrogen and oxygen. Only .... contains nitrogen. Simple sugar molecules joined together form ...., and amino acids linked up make .... A .... molecule is formed when fatty acids and glycerol combine.

2) Explain the term 'digestion' in terms of molecule size.

3) The diagram below shows the skull of a dog.

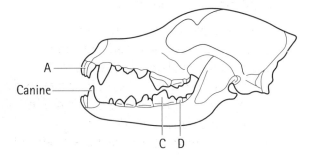

a) Which letter labels an incisor?

b) Describe the function of canine teeth in dogs.

c) A dog is a carnivore. What does this term mean?

d) Give an example of:
   i) a herbivore
   ii) an omnivore.

4) Different enzymes are responsible for the digestion of different foods.

Complete these word equations.

.... $\xrightarrow{\text{Amylase}}$ Maltose sugar

Protein $\xrightarrow{\text{.............}}$ Peptides

Fat $\xrightarrow{\text{Lipase}}$ .... + glycerol

5) The diagram below shows a human digestive system.

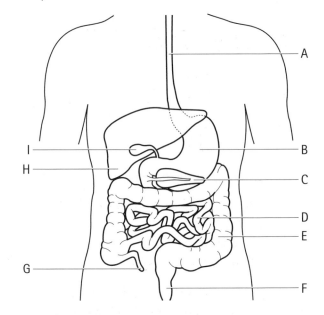

a) Which letter labels the:
   i) pancreas?
   ii) appendix?
   iii) gall bladder?
   iv) oesophagus?

b) Which letter labels the part that:
   i) stores bile?
   ii) makes pepsin?
   iii) absorbs food?

6) Complete the following sentences.

Sperm have a .... for swimming. Eggs are much larger than sperm as they contain more .... In fish, fertilisation is ....; whereas fertilisation is .... in mammals to provide .... for the sperm to swim in.

7) Explain how the number of eggs produced by fish and mammals depends on the protection provided.

8) The diagram below shows the reproductive system in a human female.

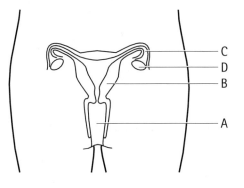

a) Name the parts labelled A, B, C and D.

b) Which letter labels the part where:
    i)   eggs are made?
    ii) fertilisation occurs?
    iii) the fetus develops?

9) From where does the embryo obtain its energy for growth and development in:
    i)   fish?
    ii) mammals?

10) List *four* ways in which a mammal loses water and *three* ways in which water is gained.

11) The diagram below shows a human excretory system.

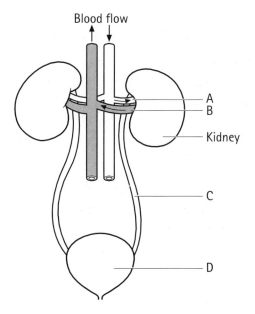

a) Name structure C.

b) What is the function of structure D?

c) Which blood vessel, A or B, is the renal artery?

12) The main waste in urine is urea.

    a) From what is urea made?

    b) Where is it manufactured?

    c) How is it transported to the kidneys for excretion?

13) Underline the correct options in this sentence describing the role of ADH in regulating water balance.

    Sweating *lowers/raises* the water concentration of the blood. The brain detects the imbalance and releases *more/less* ADH. When the ADH arrives at the kidney, it causes an *increase/decrease* in water reabsorption. As a result, a *larger/smaller* volume of urine is produced.

14) Use lines to link the correct environmental factor, animal behaviour and the advantage to the animal concerned.

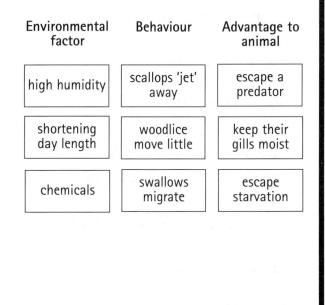

Environmental factor	Behaviour	Advantage to animal
high humidity	scallops 'jet' away	escape a predator
shortening day length	woodlice move little	keep their gills moist
chemicals	swallows migrate	escape starvation

Investigating cells

This topic is about:

- **investigating living cells**
  stains, plant and animal cells

- **investigating diffusion**
  diffusion, concentration gradients, the importance of diffusion, osmosis, the effects of osmosis on cells

- **investigating cell division**
  the function of cell division, the importance of reliable copying, the stages of mitosis

- **enzymes**
  the need for enzymes, how enzymes work, building up and breaking down, factors that affect enzyme reactions

- **aerobic respiration**
  what is aerobic respiration?; what is metabolism?

Cells are the tiny, box-like structures that form the basis of all life. By studying their activities, we can learn a lot about the unity of life, as well as finding out things that are relevant to all kinds of living organisms.

First, we will consider the cells themselves as we see them under a light microscope. All cells have a cell membrane, a nucleus and cytoplasm, whilst plant cells also have a cellulose cell wall — and possibly a vacuole, and/or chloroplasts. Many structures are more easily seen using stains.

For cells to do their huge variety of work, they need to take in lots of raw materials, and to export both products and waste materials. Fortunately, nature has provided a method that allows much of this to happen without using any energy. This is diffusion, and its special case — **osmosis**.

In order to produce the cells that are needed to replace old or damaged tissues and to allow for reproduction and growth of new organisms, cells produce identical copies of themselves. The process of **mitosis** not only produces these new cells, but also ensures the accurate copying of every piece of vital information from the old cells to the new.

Next, we consider **enzymes** — those miracle proteins that bring the reactions of cells up to unbelievable speeds at relatively low temperatures.

Finally, we take a look at the reaction that releases most of the energy we need, by combining the food we eat with the oxygen we breathe — **aerobic respiration**.

It would be helpful to know the meaning of some cell words and phrases for the exam.

Use the next few pages to find the meanings of these words and then write them beside the correct definition. Tick them off as you go.

☐ catalyst   ☐ fat   ☐ concentration gradient

☐ cell membrane   ☐ nucleus   ☐ cytoplasm

☐ carbon dioxide   ☐ chromatids   ☐ optimum

☐ osmosis   ☐ oxygen   ☐ specific

☐ mitosis   ☐ stain   ☐ phosphorylase

Words	Meanings
	controls all the activities in a cell, including cell division
	is semi-permeable and so it can control which molecules enter and leave a cell
	speeds up a chemical reaction
	is the food type that has most energy per gram
	is the difference in the concentration of a molecule on the two sides of a membrane
	are drawn to the poles of a cell by the spindle fibres
	the condition in which an enzyme is most active
	is the site of all the chemical reactions in a cell
	is released by respiring cells
	the movement of water molecules from a high water concentration to a lower water concentration through a selectively permeable membrane
	an enzyme that speeds up starch synthesis
	a coloured chemical that makes cell parts more visible
	is required for aerobic respiration
	describes an enzyme that can only speed up the reaction with one substrate
	a type of cell division

Investigating living cells

Stains

All living tissue is made of cells. These are tiny self-contained boxes, which can reproduce, grow and carry out a fantastic range of tasks. Most cells can only be seen with the help of a microscope, and even then there are problems. Much of the structure of cells is completely transparent, and this makes it difficult to see the details. Over the years, biologists have discovered a range of coloured dyes that are absorbed by various parts of the cell. These are called stains and they are used to make different parts show up clearly.

Plant and animal cells

Both animal and plant cells contain a membrane, cytoplasm and a nucleus. Plant cells also have a cell wall, and may also have chloroplasts and a sap vacuole. You should be able to recognise and identify all of these structures.

a) animal cell

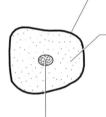

Cell membrane controls what moves in and out of the cell.

Cytoplasm (the contents of the cell). All the chemical processes happen in the cytoplasm.

The nucleus acts as a set of instructions for the cell. It's also important when cells divide because new cells inherit their characteristics from the genetic material of the nucleus.

b) plant cell

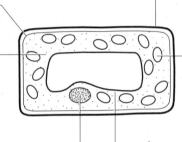

A rigid cell wall that supports the cell.

Chloroplasts, which contain chlorophyll and carry out photosynthesis.

A vacuole that contains a watery solution called sap.

Investigating diffusion

Diffusion and concentration gradients

Things that can move tend to move downhill — that's obvious. However, chemicals have a very similar habit, which we call diffusion. All substances that can move (normally we are talking about liquids or gases) always move away from higher concentrations to lower concentrations. Just as we sometimes refer to a change from a high place to a low place as a gradient, we often talk about the change from high to low concentration as a concentration gradient. Diffusion is the movement of a substance down a concentration gradient from a region of higher concentration to a region of lower concentration.

The importance of diffusion to living things

Diffusion is incredibly important to organisms because it is the process by which useful molecules enter the body cells and waste products are removed. For example, oxygen moves from a high concentration in the air sacs to a lower concentration in the blood. In the same way, digested food molecules (amino acids, glucose) move down a 'concentration gradient' from the intestine to the blood and then to any cells that have a shortage of these substances. Waste products, such as carbon dioxide or urea, do exactly the same thing in the reverse direction.

Osmosis

Osmosis is just a special case of diffusion — where water diffuses through a membrane that has different concentrations on either side. It is the movement of water through a membrane from a more watery solution towards a less watery solution.

The word 'permeable' means 'allows anything to pass through'. Membranes in cells allow small molecules (water) to pass through, but prevent larger ones from passing, so we call them 'selectively permeable'.

Pure water has the highest water concentration possible. As more salt or sugar is dissolved, the water concentration decreases. We talk about a concentration gradient when there is a high water concentration in one area and a lower water concentration in another. Now we can use these words to give us the full definition of osmosis:

Osmosis is the movement of *water molecules* through a *selectively permeable membrane* from higher water concentration to a lower water concentration — or down a concentration gradient.

> *A tank of water is divided by a selectively permeable membrane. Sea water is placed in one half of the tank, and fresh water in the other half. Predict what will happen.*

REMEMBER
When answering questions in the exam, remember that substances always move **down** a concentration gradient by diffusion or osmosis.

The effects of osmosis on cells

Animal cells are surrounded only by the membrane and may swell up and burst if too much water enters by osmosis. They just shrivel up when they lose water by osmosis.

Plant cells are very different. They have a strong cell wall outside the membrane and this wall prevents them from swelling up too much. Instead they become stiff and hard like a well-inflated football. When plant cells lose water they shrink a little, but the tough cell wall keeps its shape when the membrane inside shrinks away from it, so the cell becomes limp and floppy like a ball with not enough air in it.

a) this cell is firm because it has gained water by osmosis

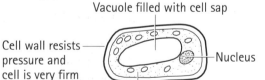

Vacuole filled with cell sap

Cell wall resists pressure and cell is very firm

Nucleus

Cytoplasm pressing on inside of cell wall

b) this cell has lost water by osmosis

Cell wall

Nucleus
Cell membrane

Cytoplasm shrunken and not pressing on cell wall, so the cell is limp

Investigating cells

Investigating cell division

📺 The function of cell division

All new cells are produced from cells that already exist by cell division. Don't ask where the very first cell came from — that's a very long story indeed! Single-celled organisms can reproduce in this way because each new cell is a complete new organism. In multicellular organisms, cell division leads to growth and development, but, in all cases, cell division is the method of increasing the number of cells. Chromosomes are thread-like structures found in the nucleus of every living cell. They carry information in the form of genes. Because it contains the chromosomes and genes, the nucleus controls all the activities of the cell (including cell division).

Chromosomes and genes

If we look carefully at the nucleus of cells under a powerful microscope, we normally only see a grainy and fairly uniform structure. But if we look at cells that are in the process of dividing, we can see tiny threads of material. These can be made even clearer if we use a suitable stain.

These thread-like structures are called chromosomes and each thread consists of a long chain of individual instructions called genes. Each gene carries information about the characteristics of the organism. The entire organism is built by following the instructions contained in them.

When a cell starts to divide, the chromosomes change themselves into a much stronger and more mobile form by coiling up to become much shorter and thicker. This makes them visible and allows them to move around in the cell to ensure that each new cell has its proper allocation of chromosomes.

The importance of reliable copying

The process of cell division is called *mitosis*. It produces two new cells that each have the same number of chromosomes as the original cell. You should be able to place drawings or descriptions of the stages of mitosis, such as the diagram on page 49, into the correct order. Each of the two cells produced by cell division has its own complete set of chromosomes and, therefore, has exactly the same information as the original cell. If the new cells had a different number of chromosomes, they would not receive the correct full information, and because this genetic information controls all the activities of the cell, it is vital that none of it is wrong or missing.

The stages of mitosis

- Just before a cell divides, each chromosome doubles up to form two identical chromatids joined by a centromere.

- The chromosomes coil up so that they shorten and thicken and become visible.

- They move to the equator of the cell and spindle fibres attach each centromere to the poles.

- The spindle fibres shorten and pull the chromatids apart.

- The chromatids are pulled to the opposite ends of the cell.

- Nuclear membranes form and the cytoplasm divides.

- There are now two new cells each with the same number of chromosomes as the original cell.

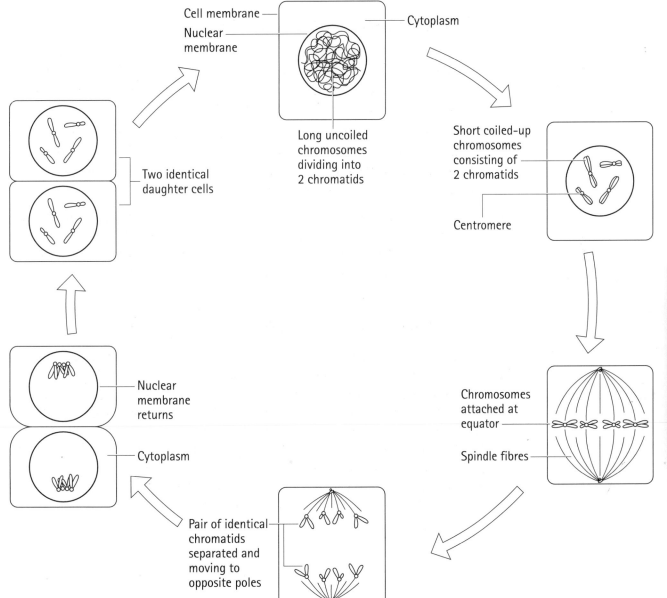

Investigating cells

Investigating enzymes

The need for enzymes

Living things work at fairly low temperatures but need many chemical reactions to keep them alive. Most chemical reactions are very slow at the temperatures found in living things. Luckily, there are special chemicals that speed up the reactions in living cells so that they are fast enough to make life possible. These are enzymes. They are all made of protein and without them the reactions in living cells would be far too slow.

A catalyst is any substance that speeds up a chemical reaction, but is left unchanged after the reaction. Enzymes are biological catalysts, and are present in all living cells.

How enzymes work

The substance on which an enzyme acts is called the substrate of that enzyme. The enzyme works by joining with the substrate, and they do this by having shapes that accurately fit together. Because each enzyme has to be shaped exactly to suit its own substrate, a different enzyme is needed for every substrate. This can be summarised by saying that enzymes are *specific*, i.e. each enzyme can only catalyse one reaction.

Building up and breaking down

! **REMEMBER** You need to know an example of a build-up reaction and an example of a breakdown reaction.

We can think of all chemical reactions as either involving the *breakdown* of a larger molecule into smaller parts (a 'breakdown reaction'), or as *building up* smaller molecules into bigger ones (a 'synthesis reaction'). A good example of a breakdown reaction is when the enzyme amylase breaks down starch into maltose (a simple sugar). An example of a synthesis reaction would be when the enzyme phosphorylase builds up a special form of glucose (glucose-1-phosphate) into the large molecules of starch.

Factors that affect enzyme reactions

- The rate of enzyme activity increases with temperature up to a maximum (about 50°C), and then falls to zero as the enzyme is denatured.

- Each enzyme has its own range of pH in which it will work.

- Optimum means 'the best' — we call the temperature or pH that makes an enzyme work fastest the optimum for that enzyme.

(?) *What do you think happens to enzyme activity when the temperature falls?*

Investigating aerobic respiration

What is aerobic respiration?

Cells need energy for various purposes – for example, cell division, movement, maintaining body temperature, and building large molecules (synthesis). Food is the source of chemical energy for most living things. Remember that fat provides more than twice as much energy per gram than either protein or carbohydrate.

The chemical process of releasing this chemical energy from the food is called respiration. By far the most efficient release of energy requires the combination of food with oxygen. Respiration that uses oxygen is called aerobic respiration.

We can describe what happens in aerobic respiration by using a word equation

glucose + oxygen → energy + carbon dioxide + water

Carbon dioxide is a waste product of respiration, which is formed from carbon and oxygen that were originally part of the food molecules.

A note about metabolism

Metabolism is a word for all the chemical reactions in a living organism. The energy released from food by respiration is important to the metabolism of all cells. Here are some of its uses:

■ to produce heat so that enzymes work more quickly

■ some reactions need energy to get them started

■ some reactions need energy to keep them going.

Investigating cells

Practice questions

1) Complete this sentence.

   .... are the basic units of living things.

2) The diagram below shows a plant cell.

   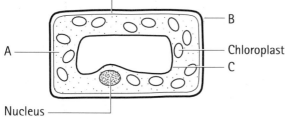

   a) Which letters label the:
      i)   cell wall?
      ii) cytoplasm?
      iii) large central vacuole?

   b) Give the function of the:
      i)   nucleus
      ii) chloroplast
      iii) cell membrane.

   c) Animal cells have only three of these structures. Which three?

3) Explain why stains are used when examining cells with a microscope.

4) Diffusion is an important process for living cells.

   a) Underline the correct option.
      Molecules move by diffusion from a *high/low* concentration to a *higher/lower* concentration *up/down* a concentration gradient.

   b) Give one example of a molecule that diffuses into a cell and one that diffuses out.

   c) Explain the importance of diffusion to living cells.

   d) Which cell structure controls the diffusion of molecules in and out of cells?

5) Osmosis is a special form of diffusion. Complete the following sentence to define osmosis.

   Osmosis is the movement of .... molecules from a .... concentration to a .... concentration through a .... membrane.

6) The diagram below shows cells that have been immersed in different solutions for 30 minutes.

   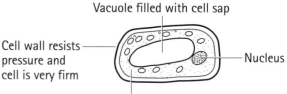

   a) Which cell, A or B, has been immersed in tap water?

   b) What has moved out of cell B?

   c) Explain why the cells were left for 30 minutes before examining them.

   d) What would be the appearance of the cells in a limp lettuce leaf – A or B?

7) Underline the correct alternative.

   When blood cells are placed in a strong salt solution, the cells *burst/shrink* because water moves *in/out* from a *high/low* water concentration to a *higher/lower* water concentration.

8) A potato chip was placed in a strong salt solution.

   a) Describe how it changes over the next hour.

   b) What improvement could be made to increase the reliability of the results from this experiment?

9) The diagrams below show cells in various stages of cell division. The sequence has been mixed up.

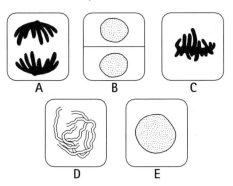

   a) What is the alternative word for cell division?

   b) Organise the letters to show the correct order of the stages.

   c) Describe the next event after the stage shown in diagram C.

   d) Explain why it is important for the number of chromosomes in a daughter cell to be the same as in the parent cell.

   e) Which part of the cell controls cell division?

10) When a lump of liver was placed in a test tube with hydrogen peroxide, many gas bubbles were seen to rise to the surface.

   This shows that the liver contained an .... , which acted as a biological .... because it increased the rate of a chemical reaction. A control for this experiment would have everything exactly the same, except there would be no

   ....

11) a) Give an example of an enzyme that breaks large molecules into small.

    b) Name an enzyme that synthesises large molecules from small.

    c) From what are all enzymes made?

12) The graph below shows the effect of temperature on the activity of the enzyme amylase.

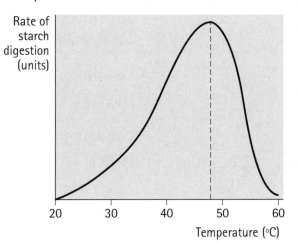

   a) Describe the effect of temperature on the activity of amylase.

   b) What term is used to describe the temperature at which amylase is most active.

   c) Explain why the starch would not have been broken down if pepsin had been used instead of amylase.

13) Which type of food, protein, fat or carbohydrate, contains the most energy per gram?

14) Complete the following sentences.

   The most important product released during aerobic respiration is .... For food to be broken down completely .... gas must be present. Water and .... gas are also made.

The body in action

This topic is about:

- **movement**
  the functions of the skeleton, joints, bones, tendons and muscles

- **the need for energy**
  energy, breathing, gas exchange, features of efficient exchange structures, the heart, blood circulation, types of blood vessel, components of blood, gas exchange and the blood

- **co-ordination**
  the eyes, the ears, the nervous system, reflex actions, the brain

- **changing levels of performance**
  muscle fatigue, fitness and exercise

This topic introduces some of the biological principles involved in the study of human movement and physical performance. It also examines the relationships between physical activity and healthy living.

The skeleton is the framework that supports us, and makes movement possible. At the same time, many of the bones provide protection for delicate organs.

Movement requires energy, and this energy needs to be readily available to all parts of the body. This demands the effective co-operation of several systems. The digestive system (covered in the Animal Survival topic, page 30) gathers dissolved nutrients. The breathing system gathers oxygen and disposes of carbon dioxide. Connecting the whole thing together is the blood system, which transports raw materials from the lungs and digestive system to the cells that will use them, and waste materials to the site of their excretion. The heart pumps to maintain circulation, and each

different type of blood vessel is ideally adapted to its particular function.

Movement would, however, be a dangerous and often pointless exercise if we could not control it and relate it to our surroundings. The next sub-topic, therefore, considers the ways in which we detect our environment, and especially the senses of sight and hearing. The nerves serve to connect our senses to the brain, and our brain to our muscles. At the centre of all our activities lies the brain, which carries out the unimaginably complex task of co-ordinating all incoming and outgoing information.

We end the topic by considering some of the ways in which we can help the complex instrument that is our body to function at its best. By training and exercise, it is possible for us not only to improve performance, but to reduce the damage and strain caused by the demands of a busy and active life.

It would be helpful to know the meaning of some body words and phrases for the exam.

Use the next few pages to find the meanings of these words and then write them beside the correct definition. Tick them off as you go.

- ☐ skull
- ☐ ball and socket
- ☐ synovial fluid
- ☐ tendon
- ☐ diaphragm
- ☐ cilia
- ☐ left ventricle
- ☐ coronary artery
- ☐ retina
- ☐ auditory nerve
- ☐ reflex action
- ☐ cerebrum
- ☐ fatigued
- ☐ lactic acid
- ☐ recovery time

Words	Meanings
	connects muscle to bone
	is a sore and inefficient muscle after repeated contraction
	is the part of the skeleton that protects the brain
	is a fast response to danger using a relay neurone to connect the sensory neurone to the motor neurone
	converts light energy into electrical energy
	contracts for breathing in
	is a heart chamber that pumps the blood around the body
	is a type of joint that allows movement in three planes
	is a chemical produced by anaerobic respiration in muscles
	is the part of the brain where thought and memory occur
	is an oily liquid that lubricates a joint to reduce friction
	carries electrical impulses from the cochlea to the brain
	sweeps mucus and dust out of the lungs
	is the period of time between the end of exercise and when the heart rate returns to normal
	provides the heart muscle with food and oxygen

Movement

The functions of the skeleton

Your skeleton is important for *support*, *movement* and *protection*. It provides a framework for support and for attaching muscles. The protection of the skeleton is particularly important for the brain, the heart and lungs, and the spinal cord.

Write down the names of the parts of the skeleton that protect each of these parts of the body.

Joints

Because bones have to be solid and strong for support and protection, the skeleton needs joints between the bones to allow for movement. You need to know about two kinds of joint: the ball and socket joint, and the hinge joint.

- *Ball and socket* joints are found at the shoulder and the hip — they are able to move in all three dimensions.

- *Hinge joints* are found at the elbow and knee — they allow movement in one plane only.

You should know the following parts of the joint and the function of each part.

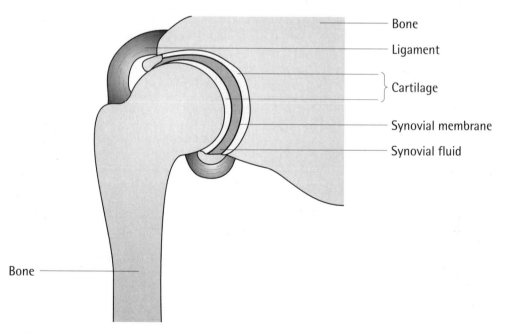

- The ligaments hold the bones together.

- The synovial membrane makes synovial fluid.

- Synovial fluid is oily and acts as a lubricant.

- The cartilage is a soft, smooth substance that allows smooth movement as well as acting as a shock absorber.

Bones, tendons and muscles

- *Bone* is made of tough, flexible fibres and hard minerals. Bones are built of living cells, so they need a blood supply for food and energy.

- *Tendons* are the tough, non-stretchy structures that attach muscles to bones. They need to avoid stretching to make sure that all the movement of the muscle is passed on to the bone.

- *Muscles* move joints and limbs when they contract. The contraction makes the tendon pull on the bone, which makes it move. Since muscles can't push but can only pull bones, every joint needs a pair of opposing muscles to make it work. When one of the muscles contracts the other one relaxes. One muscle in the pair contracts to bend the joint; the other contracts to straighten the joint.

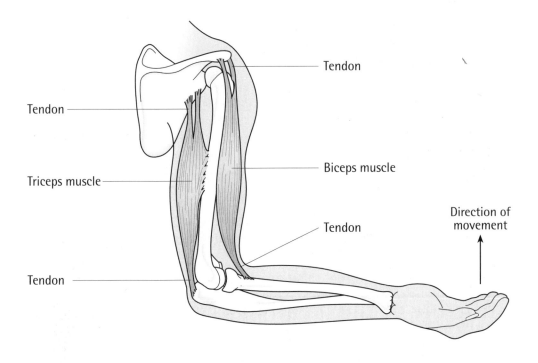

The diagram shows the structures involved in bending and straightening the human elbow. The tendons attach the muscles to the bones.

(?) *Which muscle contracts to bend the elbow joint?*

The need for energy

Energy

Energy is used all the time in living cells and organisms. This energy is provided in the form of chemical energy in the food we eat. The energy is released from food by a process called *respiration*.

The amount of energy you need depends upon your age, sex and your level of activity. If you eat food with more energy than you use, your body will store the excess as body fat – you will gain weight. On the other hand, if you eat food with less energy than you require your body will use up stored fat – you will lose weight.

(?) *Place the following people in order of how much energy they are likely to need: a child, a computer operator, an athlete, an elderly person.*

Breathing

Several important chemicals are used and released during the process of releasing the energy from our food. Oxygen is needed if the body is to release energy efficiently. Carbon dioxide and water are waste products of respiration. During breathing, the lungs absorb the oxygen we need and release waste carbon dioxide. We also release a lot of water vapour from our bodies as we breathe out.

When we breathe, our ribs and diaphragm move to increase and decrease the volume of the thorax. This makes the lungs expand to fill with fresh air and then contract to breathe out.

(!) REMEMBER The ribs move up and out to increase the chest volume when we breathe in; the diaphragm pulls down.

(!) REMEMBER The ribs move down and in when we breathe out whilst the diaphragm relaxes and moves upwards.

Structure of the breathing system

You should be familiar with all the parts shown on the diagram of the lungs (page 59, top). As air enters through the nose and mouth it first passes through the *windpipe* (trachea), which is reinforced by *cartilage rings* that help to keep it open. The windpipe branches into two *bronchi* (one to each lung), which in turn split into smaller and smaller tubes called *bronchioles*. The bronchioles end in microscopic *air sacs* (alveoli), which are lined with *mucus* and surrounded by *blood* capillaries.

The mucus is needed to keep the inside of the lungs moist so that the gases can dissolve and diffuse into (or out of) the blood. Mucus is sticky so that it covers all the inner surfaces of the lungs without trickling downwards. The sticky mucus also traps any dust and micro-organisms that are breathed in.

Tiny hairs called cilia cover all the cells lining the air tubes, and these move back and forth to sweep the mucus upwards towards the throat. This helps to remove the dust and micro-organisms, and is important in protecting the lungs from damage.

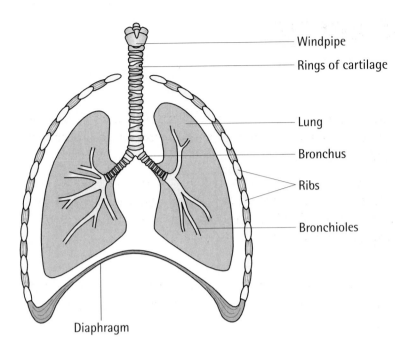

- Windpipe
- Rings of cartilage
- Lung
- Bronchus
- Ribs
- Bronchioles
- Diaphragm

Gas exchange

The air sacs (see right) have a very large total surface area and a very good blood supply, which makes them very efficient at exchanging gases with their surrounding blood vessels (capillaries). Oxygen diffuses from the air sac into the blood; carbon dioxide diffuses from the blood into the air sac.

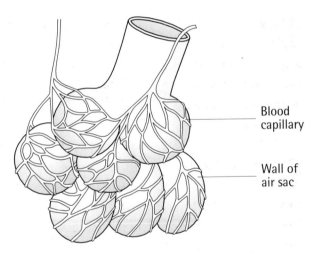

- Blood capillary
- Wall of air sac

Features of efficient exchange surfaces

Three features are particularly important in our lungs. Most of them also apply to other exchange surfaces such as intestines, leaves etc.

- *A good blood supply* — the air sacs have a large capillary network so that large volumes of gases can be exchanged.

- *Thin* — the air sacs are very thin so that gases can easily diffuse through them.

- *Moisture* — the air sacs are moist with mucus so that gases can dissolve.

- *Large surface area* — the area for gases to diffuse through in human lungs is roughly the area of a tennis court.

(?) *What factors might speed up your breathing rate?*

The body in action

📺 The heart

The heart is a tireless and powerful pump that moves your blood in a constant flow round your body. One half collects de-oxygenated blood from the tissues and sends it off to the lungs for oxygen; the other, more powerful half collects the newly oxygenated blood from the lungs and sends it round the rest of the body.

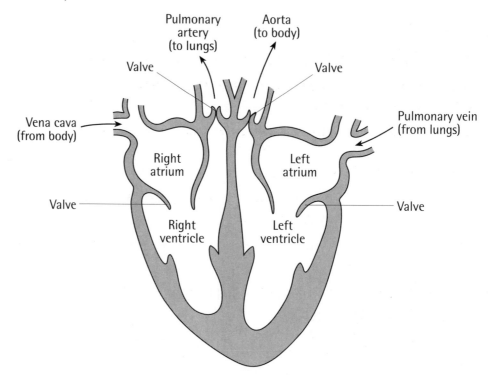

REMEMBER
You need to know the names of the heart chambers and the blood vessels connected to each.

You must be able to identify the parts of the heart shown on the diagram above – the chambers, blood vessels and valves. Next, you need to be familiar with the path followed by blood as it passes through the heart and its blood vessels.

> body → vena cava → right atrium → right ventricle → pulmonary artery → lungs
>
> lungs → pulmonary vein → left atrium → left ventricle → aorta → body

Finally, you should be aware of the following important facts about the heart:

- the valves prevent blood from flowing backwards through the heart
- the left ventricle is much thicker than the right because it has to pump blood round the whole body – the right ventricle is thinner because it only has to pump blood through the lungs
- the heart muscle gets its own blood supply from the coronary arteries.

Blood circulation

There are three main types of blood vessel that contain the blood as it circulates round the body. Each type has a different structure, and a different job to do — in many ways, your circulation system is like a road transport system with motorways, main roads and side roads all playing an important part in keeping the traffic moving.

Types of blood vessel

- Arteries take blood away from the heart to the body organs and tissues.
- Capillaries are tiny, thin-walled vessels that form a network to take blood through the organs and tissues.
- Veins collect blood from the capillaries and return it to the heart.

The components of blood

- Blood is a pale yellow liquid called plasma with red blood cells, white blood cells and platelets suspended in it.
- Plasma is mainly water with lots of substances, such as glucose, salts, hormones and amino acids, dissolved in it.
- Red blood cells collect oxygen in the lungs and carry it round the body.
- White blood cells fight infections.
- Platelets help to clot the blood where there has been a cut.

Red blood cell are disc-shaped and dented in the middle.

White blood cells contain nuclei.

Plasma

Platelets

Gas exchange and the blood

All cells use up oxygen to release energy. This oxygen comes from the red blood cells in the nearby capillaries, by diffusion. In the opposite direction, the carbon dioxide, which is produced as a waste product by the cells, diffuses from the cells into the blood plasma. Red blood cells contain a dark red chemical called haemoglobin, which combines with oxygen in the lungs to form a bright red chemical called oxyhaemoglobin. When the blood gets to places where oxygen is being used up, oxyhaemoglobin releases the oxygen and turns back into haemoglobin.

The capillary networks in the body are very well adapted for exchanging substances with the body cells. This is because capillaries have very thin walls that allow gases to diffuse in or out, and they pass between the cells of every tissue in very large numbers to ensure that every cell has capillaries passing very close to it.

The body in action

Co-ordination

The eyes

You need to know the basic structure of the eye and the function of the parts, as shown in the diagram below, together with a couple of simple eye facts.

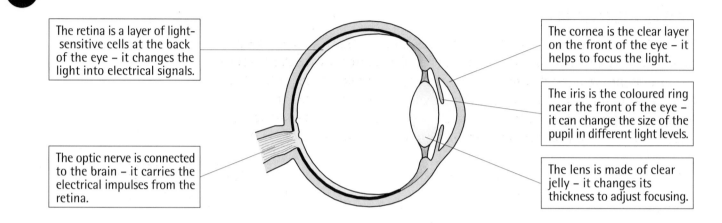

The retina is a layer of light-sensitive cells at the back of the eye – it changes the light into electrical signals.

The optic nerve is connected to the brain – it carries the electrical impulses from the retina.

The cornea is the clear layer on the front of the eye – it helps to focus the light.

The iris is the coloured ring near the front of the eye – it can change the size of the pupil in different light levels.

The lens is made of clear jelly – it changes its thickness to adjust focusing.

Eye facts

- Having two eyes rather than one is called binocular vision. This makes judgement of distances more accurate.

- Each eye sees a slightly different picture and the brain puts the two views together to give a three-dimensional image.

The ears

You need to know the basic structure of the ear and the function of the parts, as shown in the diagram below. Apart from this, you need to know only one ear fact: having two ears rather than one makes judging the direction of sounds more accurate.

Structure of the human ear

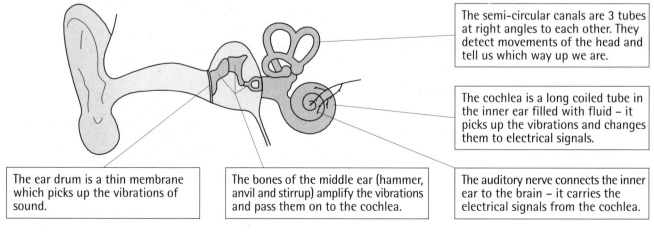

The semi-circular canals are 3 tubes at right angles to each other. They detect movements of the head and tell us which way up we are.

The cochlea is a long coiled tube in the inner ear filled with fluid – it picks up the vibrations and changes them to electrical signals.

The ear drum is a thin membrane which picks up the vibrations of sound.

The bones of the middle ear (hammer, anvil and stirrup) amplify the vibrations and pass them on to the cochlea.

The auditory nerve connects the inner ear to the brain – it carries the electrical signals from the cochlea.

The nervous system

■ The nervous system is made up of the *brain*, the *spinal cord* and *nerves*.

■ The brain and spinal cord are the *central nervous system*.

■ Nerves carry information from the senses to the central nervous system and from the central nervous system to the muscles.

Reflex actions

Normally, information is taken to the brain, and instructions come from the brain, but in a reflex action the brain may be missed out for the sake of speed. Reflex actions are a rapid automatic response to a stimulus. They protect the body by reacting quickly to danger. The speed is possible because the information travels directly from the sense organs to the muscles through a direct link in the central nervous system, i.e. not immediately via the brain.

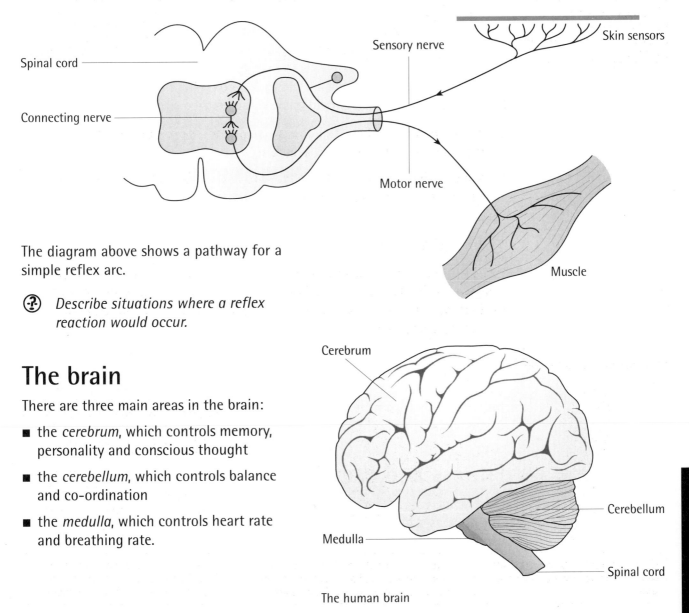

Spinal cord

Connecting nerve

Sensory nerve

Skin sensors

Motor nerve

Muscle

The diagram above shows a pathway for a simple reflex arc.

(?) *Describe situations where a reflex reaction would occur.*

The brain

There are three main areas in the brain:

■ the *cerebrum*, which controls memory, personality and conscious thought

■ the *cerebellum*, which controls balance and co-ordination

■ the *medulla*, which controls heart rate and breathing rate.

Cerebrum

Cerebellum

Medulla

Spinal cord

The human brain

Muscle fatigue

If you make any of your muscles work very hard or for a very long time, they begin to ache and stop working as well. In extreme cases, they may even stop working altogether for a while. This is called muscle fatigue. It is caused by a build-up of a substance called *lactic acid* in the muscles.

When we make our muscles move, they need energy. This is provided by respiration. This is the process of combining food, especially glucose, with oxygen, to release energy, carbon dioxide and water. If the need for energy is particularly high, such as in a sprint, or if an unfit person takes some unaccustomed exercise, then it sometimes happens that the muscles are using up energy so fast that the blood is not able to bring oxygen to the cells quickly enough to supply the demand.

It would obviously be a potential disaster if the muscles simply refused to work when this happens. The individual might be escaping from danger, or might be a starving predator needing some last few steps to catch a juicy meal. Luckily, the cells of the muscles have a back-up system that can come into action as a short-term help in providing extra energy for urgent demands. This is called *anaerobic respiration*, which is just a fancy way of saying respiration without oxygen.

Anaerobic respiration

When there is a shortage of oxygen in the muscle cells, they can release some of the energy from glucose without it. This is nowhere near as efficient as respiration with plenty of oxygen (aerobic respiration), but can make all the difference in an emergency. The main drawbacks are that much less energy is released during anaerobic respiration, and that a poisonous waste product is made – *lactic acid*. In the muscles, lactic acid causes pain and stiffness, so the use of anaerobic respiration can only be for a limited time. Fortunately, after exercise, lactic acid is quickly broken down using oxygen, and the muscles soon recover. The oxygen needed to get rid of the lactic acid explains why we are still breathless for a while after strenuous exercise. We are providing the oxygen that we were short of during the sprint – we are repaying the *oxygen debt*.

Fitness and exercise

Exercise increases the rate at which energy is needed from food, and this, in turn, increases the need for both food and oxygen in the body. This is why your pulse rate and breathing rate increase with exercise. Your pulse is just an indication of your heart rate, and your heart speeds up to pump extra food and oxygen to the muscles. At the same time, breathing speeds up to get more oxygen and to get rid of waste carbon dioxide.

When a fit person (such as an athlete) exercises, the pulse rate, breathing rate and lactic acid levels rise much less than they do in an unfit person. The time that it takes for pulse and breathing rate to return to normal is called the *recovery time*, and the fitter you are, the shorter your recovery time.

The effects of fitness training

Training allows a person to exercise more vigorously and for a much longer time before muscle fatigue sets in. Fitness training improves your body's efficiency in several ways:

- it makes your heart able to *pump more blood* every beat
- it *increases the flow of blood* through the muscles
- it *increases lung volume*, which means that each breath carries more oxygen.

All of these provide food and oxygen to the muscles much faster – so your pulse rate can stay lower. Because much more oxygen reaches the muscles, less lactic acid is made.

Recovery time

Your recovery time is really a measure of the time it takes your body to repay the oxygen debt in your muscles. If you are fit, you make less lactic acid so there is less to get rid of. An efficient heart and lungs provide the oxygen much more quickly.

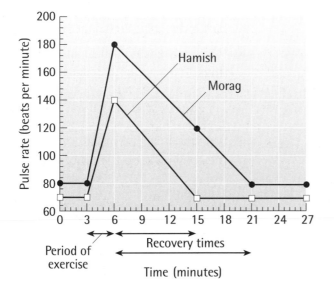

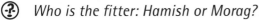

? *Who is the fitter: Hamish or Morag?*

Practice questions

1) The diagram below shows a human skeleton

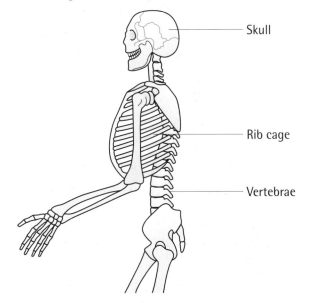

Skull

Rib cage

Vertebrae

a) One of the functions of the skeleton is protection. What is protected by the

   i) skull   ii) rib cage   iii) vertebrae?

b) State two other functions of the skeleton.

2) The diagram below shows the structure of the human arm.

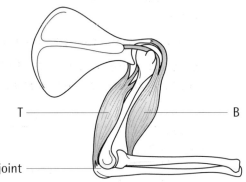

T

B

Elbow joint

a) Which type of joint is the elbow and what kind of movement does it allow?

b) Name the structure that attaches muscle to bone.

c) Describe how muscles T and B straighten the arm.

3) A roasted bone is hard but crumbles. Which bone component has been burnt away?

4) Underline the correct alternative.

People with an energy input that is *less/more* than their energy output lose weight.

5) The diagram below shows the structure of the human breathing system.

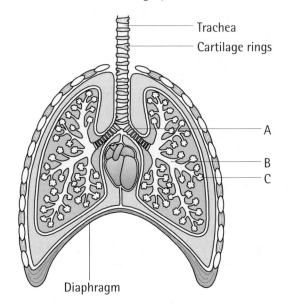

Trachea

Cartilage rings

A

B

C

Diaphragm

a) Name the structures labelled A, B and C.

b) What is the function of the cartilage rings?

c) Explain how the diaphragm causes breathing in.

d) Describe how the mucus and cilia lining the trachea keep the lungs clean.

6) This diagram shows the heart and main blood vessels.

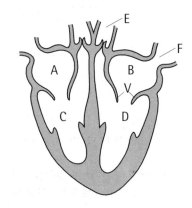

E

F

A

B

V

C

D

a) Name chamber B and blood vessel E.

b) From where has vessel F brought blood?

c) Explain why the walls of chamber D are thicker than chamber C.

d) What is the function of structure V?

e) Name the blood vessel that provides the heart with its own blood supply.

7) The answer to the following questions is either artery, capillary or vein.

Which type of blood vessel:
a) has valves?
b) has walls one cell thick?
c) carries blood away from the heart?
d) has a pulse?
e) exchanges gas with cells?
f) carries blood back to the heart?

8) a) Name one molecule carried by a red blood cell, and one carried by the plasma.

b) What is the name of the red pigment carried in red blood cells?

9) The following diagram shows the structure of the human eye.

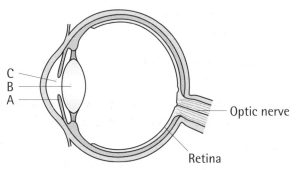

a) Name the structures labelled A, B and C.

b) Give the function of the retina and optic nerve.

c) Explain the advantage of having two eyes looking forward at the same object.

10) Explain how the arrangement of the semi-circular canals is related to their function.

11) The diagram below shows the structure of the human brain.

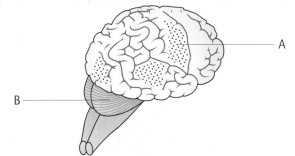

a) Name the parts labelled A and B.

b) Describe the function of the medulla.

12) Rapidly contracting muscles become short of oxygen and have to rely on anaerobic respiration.

a) What term is used for sore and inefficient muscles after strenuous exercise?

b) Complete the following word equation for anaerobic respiration in muscles.

food → .... + energy

13) The diagram below shows the effect of exercise on the heart rate of two girls.

a) What was the heart rate of Morag before exercise?

b) What does the term 'recovery time' mean?

c) Give two pieces of evidence that suggest that Isobel is fitter than Morag.

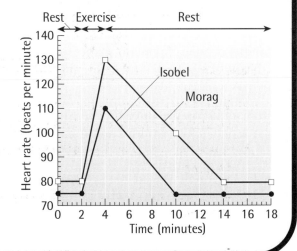

Inheritance

This topic is about:

- **variation**
  the concept of species, continuous variation, discontinuous variation

- **what is inheritance?**
  genetic information, chromosomes and genes, genotypes and phenotypes, genetic crosses, the difference between expected and actual results, sex determination

- **genetics and society**
  selective breeding, mutations, factors that cause mutations, amniocentesis

The study of inheritance has enormous social and economic importance. Knowledge of the processes of plant and animal breeding has brought about revolutions in agriculture.

Organisms have been produced that have become more and more specialised to give bigger and better yields. Man is rapidly approaching the point where genetic manipulation has the potential to predict and control changes in organisms.

The starting point is the simple observation of differences between individual organisms. This variation is the reservoir of genetic information. It has generated the stability and adaptability of life on our planet, and it offers humans the potential to adapt and control it for our own survival and benefit.

From variation we move to a consideration of the way in which information is passed from generation to generation. The way in which the information is mixed during sexual reproduction in order to produce endless new combinations of characteristics allows us to understand and, to a limited extent, predict the inheritance of the particular aspects of organisms that make them individual.

Finally, we look at some of the ways in which the science of inheritance affects our lives as human beings. The occurrence and consequences of mutations are considered, together with an example of how to detect particular instances. The skills and benefits of selective breeding have allowed man to move from a nomadic hunter-gatherer existence, to a static and highly effective agricultural society that has almost kept pace with the vastly increasing demands for food over the centuries.

FactZONE

It would be helpful to know the meaning of some inheritance words and phrases for the exam.

Use the next few pages to find the meanings of these words and then write them beside the correct definition. Tick them off as you go.

- [ ] F_1
- [ ] X or Y
- [ ] Down's syndrome
- [ ] species
- [ ] genotype
- [ ] alleles
- [ ] phenotype
- [ ] gametes
- [ ] continuous variation
- [ ] selective breeding
- [ ] amniocentesis
- [ ] true breeding
- [ ] mutation factor
- [ ] dominant
- [ ] discontinuous variation

Words	Meanings
	chemicals or X-rays or UV light
	is the appearance of an organism for a particular characteristic
	are the different forms of a gene
	is the first generation in a genetic cross
	is a technique that can be used to detect chromosome mutations before birth
	are the sex chromosomes carried by a sperm
	is the collective name for sperm, eggs, ovules and pollen
	organisms which have the same phenotype generation after generation because both their genes are the same
	has a wide range of values such as height and weight
	is a human condition resulting from a chromosome mutation
	is the gene whose effect always shows in the phenotype
	is a group of interbreeding organisms who have fertile offspring
	examples are eye colour, blood groups and sex
	is when only the fastest pigeons are allowed to breed
	is the genes an organism has for a particular characteristic

Variation

A species

A species is a group of organisms that can breed with each other and produce fertile offspring.

The first part is easy to understand. For example, a female dog can mate with any male dog and have puppies. But what about the 'fertile offspring' part? Well, this detail overcomes the fact that closely related species can sometimes mate and produce young. A well-known example is the mule, which is the result of a mating between a member of the horse species and a member of the donkey species. Two different species! However, two mules cannot produce a baby mule because they are infertile.

So for two animals to be a member of the same species they must be able to produce young and also their offspring must be able to produce young themselves. It is the same for plants.

Variation

Variation is the differences that exist between members of the same species. The picture below shows just some of the hundred or so horse breeds.

From the towering Clydesdale to the tiny Falabella they are all part of one species. As well as size, they differ in many other ways, which include shape, colour, speed and temperament.

Every horse is a unique individual and this is also true of members of the human, oak or fiddler crab species.

(?) *Can you think of several other animals that humans have bred to create greater variation within a species?*

Types of variation

There are two kinds of variation — *continuous* and *discontinuous*. You may be asked for examples of each.

Continuous variation

When asked to give the meaning of continuous variation you should write 'there is a wide range of values between two extremes'.

This histogram shows the pulse rates for 200 adults. See how pulse rates fit the definition of continuous variation as they vary from between 50 beats per minute up to 94 beats per minute.

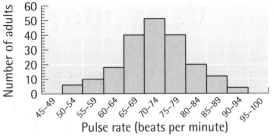

(?) *Can you think of other examples from everyday life of continuous variation?*

Discontinuous variation

When asked to give the meaning of discontinuous variation, you should write 'there are distinct groups with no values in between'.

The bar graph shows the blood groups for 200 adults. As you can see, blood groups fit the definition of discontinuous variation as the adults are divided into four separate blood groups with nothing in between.

(TV) *Watch the video sequence for more examples of variation.*

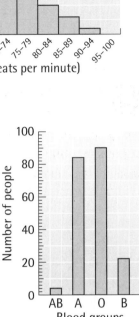

Some more examples of variation

(?) *Decide which of the following examples are continuous variation and which are discontinuous variation.*

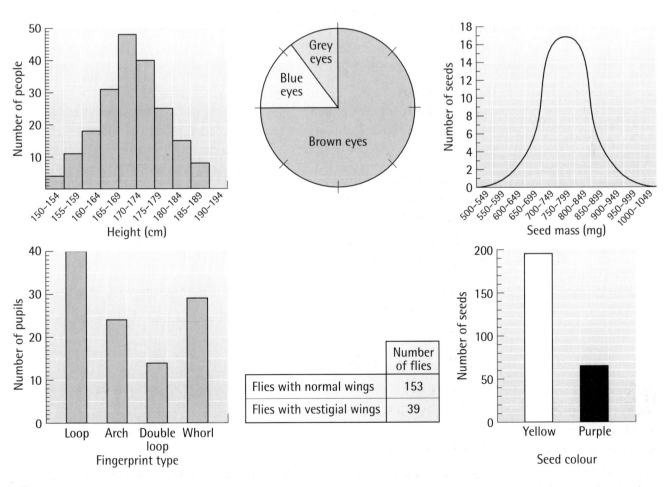

(◎) *Check your understanding with the Test bite.*

What is inheritance?

Genetic information

Most characteristics of plants and animals are inherited from their parents.

Scorpion   Buttercup

The reason a scorpion and a buttercup plant are so different is because their parents gave them different genes. The scorpion was given genes for legs, pincers and a stinging tail, whereas the buttercup plant was given genes for flowers, leaves and roots.

(?) *List some of the characteristics your parents passed on to you.*

Chromosomes and genes

Half of the genes that made you what you are today were passed on to you in a set of 23 chromosomes in the sperm from your father. These were combined with a set of 23 chromosomes in the egg from your mother. This gives you two matching sets totalling 46 chromosomes.

But why if everyone has 46 chromosomes in each body cell do sperm and eggs, or gametes as they are collectively known, have only 23 chromosomes? The halving of the number of chromosomes is because of a special kind of cell division that happens only in the ovaries and testes. Thus gametes only have a single set of chromosomes. At fertilisation, the two sets merge to give your first cell a double set.

! **REMEMBER** Sperm, eggs, pollen and ovules are collectively known as gametes.

The following diagram summarises the whole process.

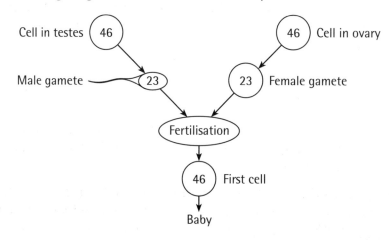

! **REMEMBER** There are two sets of chromosomes in a body cell but only a single set in a gamete.

(?) *If an onion epidermis cell has 16 chromosomes, how many will be in an onion pollen grain?*

BITESIZEbiology

Phenotypes

The phenotype of an organism is what it looks like for a particular characteristic.

Most characteristics have at least two phenotypes as there is usually more than one form of a gene. Different forms of the same gene are *alleles*.

A pea plant may be tall or dwarf and also may have white or purple flowers. You may, or may not, be able to roll your tongue. The picture shows fruit flies that may have either normal or curly wings. These are all examples of a characteristic having two or more phenotypes.

Normal and curly winged
Drosophila

True breeding

If two organisms with the same phenotype are crossed and all their offspring have the same phenotype as their parents they are said to be true breeding.

But what if two true breeding organisms with different phenotypes are crossed? The pictures below show a cross between two rabbits with different phenotypes. The parents, or P generation, are different. One is a true breeding albino; the other a true breeding Himalayan.

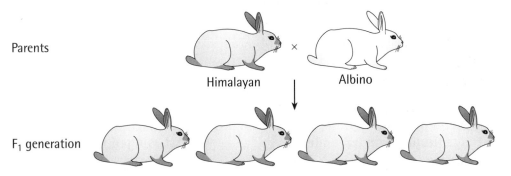

Parents

Himalayan × Albino

F_1 generation

All the offspring, or the F_1 generation, are the same. They all show the Himalayan characteristic although they must also have an albino gene from the other parent. The Himalayan gene from one parent hides the effect of the albino gene from the other. The Himalayan gene is said to be dominant and albino is a recessive gene.

Genotype

These are the genes an organism has for a particular characteristic. Each characteristic is controlled by a pair of genes, one carried by the male gamete and one by the female gamete.

We use letters as a shorthand way to describe the genes. A capital letter for dominant alleles and a small letter for recessive alleles. A true breeding animal for a dominant characteristic might be BB; whereas bb is a true breeding animal with the recessive characteristic. A non-true-breeding animal will have one dominant and one recessive allele and so will be Bb.

! REMEMBER
In the exam, you may be asked to decide which gene is dominant by looking at the offspring from a cross.

A genetic cross

Here is an example of how a cross involving one pair of genes — a monohybrid cross — is worked through to the second, or F_2, generation.

In the fruit fly *Drosophila*, normal wings (N) are dominant to vestigial (n).

◉ *Show the F2 results expected from a cross between a true breeding normal-winged male and a true breeding vestigial-winged female.*

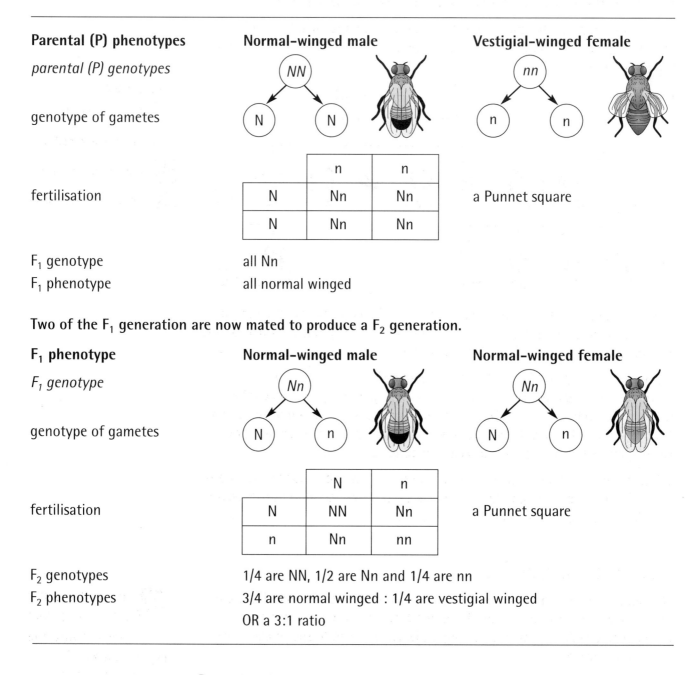

Parental (P) phenotypes	Normal-winged male	Vestigial-winged female

parental (P) genotypes — NN — nn

genotype of gametes — N, N — n, n

fertilisation — a Punnet square

	n	n
N	Nn	Nn
N	Nn	Nn

F_1 genotype — all Nn
F_1 phenotype — all normal winged

Two of the F_1 generation are now mated to produce a F_2 generation.

F_1 phenotype	Normal-winged male	Normal-winged female

F_1 genotype — Nn — Nn

genotype of gametes — N, n — N, n

fertilisation — a Punnet square

	N	n
N	NN	Nn
n	Nn	nn

F_2 genotypes — 1/4 are NN, 1/2 are Nn and 1/4 are nn
F_2 phenotypes — 3/4 are normal winged : 1/4 are vestigial winged
OR a 3:1 ratio

 Watch the video sequence of a cross between white- and purple-flowered pea plants.

Expected and actual results

The actual numbers of offspring of each type produced in the F_2 is unlikely to be exactly as predicted. Only one of the two alleles can go into a gamete, and this is random. It is also a matter of chance which gametes combine to form the offspring. You get much nearer the expected result if you deal with a large number of offspring because this minimises the effects of random variations.

REMEMBER
You may be asked why the expected and actual results from a genetics cross are not the same.

Working out genotypes in a family tree

Here is an example. The ability to roll the tongue is controlled by two forms of the same gene – two alleles. The family tree below shows the inheritance of tongue rolling.

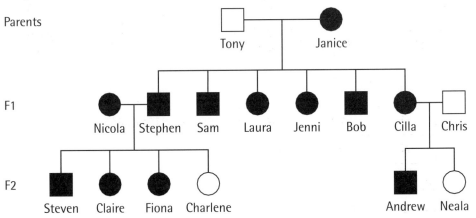

Key to symbols

☐ Non-rolling males

○ Non-rolling females

■ Tongue-rolling males

● Tongue-rolling females

Most people in the family are tongue rollers. This suggests that the tongue-rolling gene is dominant. For example, Tony and Janice produced only tongue-rolling children although Tony is a non-roller.

Who is true breeding? Certainly, *all* those who show the **recessive characteristic** must be; if they had a dominant gene they would be rollers. Most of the rollers are probably not true breeding. Janice is almost certainly true breeding with two dominant genes as all six of her children are tongue rollers. She seems not to have a recessive gene to pass on.

What is the genotype of Nicola and Stephen? You can see Charlene is a non-roller. She must have been given a recessive gene by each of her parents. So they both must be Rr. Similarly, Cilla must be Rr.

Sex determination

The sex of a child is determined by the sex chromosomes. Females have two X chromosomes (XX), and males have one X chromosome and one Y chromosome (XY). This means that eggs always carry an X: sperm can have either an X or a Y.

When an egg (X) joins with a (Y) sperm, the result is a boy (XY). When an egg (X) joins with an (X) sperm, the result is a girl (XX).

 Log on to see the sex determination movie.

REMEMBER
XX = girl and XY = boy – "Only the guys get the Ys".

Genetics and society

Selective breeding

Animals and plants have many characteristics that are useful to man. From the earliest times man has been picking his best plants and animals for breeding. Over many generations, the organisms became better and better for human use.

The picture shows the wild ox from which modern-day cattle originated.

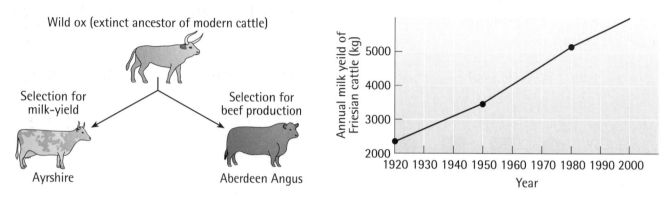

Wild ox (extinct ancestor of modern cattle)

Selection for milk-yield

Ayrshire

Selection for beef production

Aberdeen Angus

REMEMBER You should be able to give at least one plant or one animal example of an improved characteristic resulting from selective breeding.

During this century, stockmen have been selecting for either meat production or high milk yields. This has resulted in specialist breeds such as Aberdeen Angus for beef and Ayrshires for milk. As the graph shows, selection has been particularly successful in increasing milk yields. Even the domesticated animals we do not eat are bred selectively. For example, dogs have been bred for different purposes such as guard dogs, sheepdogs, tracker dogs and 'cute pets'.

The picture below shows how selection by man has produced many vegetables from the wild sea cabbage, by selecting for particular characteristics and then, generation after generation, using the plants judged to be the best for breeding. As usual, the rest were eaten.

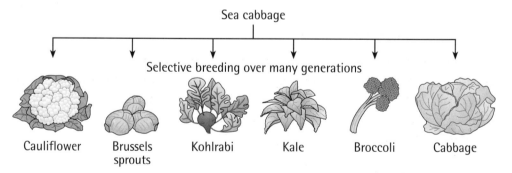

Sea cabbage

Selective breeding over many generations

Cauliflower | Brussels sprouts | Kohlrabi | Kale | Broccoli | Cabbage

There are many other examples of selection in plant breeding. If the plant can be eaten, the breeder will have been selecting for high yields and resistance to diseases. In the exam, you could use almost any kind of edible plant as an example.

(?) *Can you think of another species of plant where several different varieties have been produced by selective breeding?*

Chromosome mutations

Sometimes, chromosomes can change. These changes can occur quite naturally, and are called mutations. A mutation can cause changes in characteristics.

REMEMBER Be able to give an example of a chromosome mutation of economic importance.

Down's syndrome

An example of a human chromosome mutation is when a child inherits an extra copy of a particular chromosome from its mother. Each cell has three chromosomes instead of two, and this causes a condition called Down's syndrome.

The picture shows the chromosomes from a person with Down's syndrome.

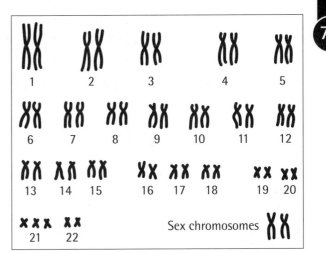

(?) *A person with Down's syndrome has an extra copy of chromosome number ....*

(?) *The sex of the person whose chromosomes are shown in the picture is ....*

Amniocentesis

It is possible to examine the chromosomes of a human embryo at a very early stage of pregnancy to check for signs of Down's syndrome using a process called amniocentesis. The picture shows that amniocentesis involves removing a small volume of the amniotic fluid that surrounds and protects the embryo. The fluid always contains a few cells from the embryo. These cells can be examined under a microscope to find out how many chromosomes they contain.

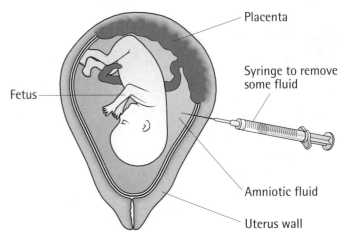

Useful chromosome mutations

Not all chromosome mutations are harmful. In many crop plants, such as wheat or barley, the cells have complete extra sets of chromosomes because of a chromosome mutation. This gives them extra disease resistance and hardiness. Most modern crop plants have multiple sets of chromosomes, which make them more profitable than the older varieties.

Mutagenic agents

The frequency of mutations happening in a population can be increased if they are exposed to certain factors. Various types of chemicals, extremes of temperature and radiation, such as X-rays and ultraviolet light from the sun, can speed up the rate of mutation.

REMEMBER You should be able to give an example of a factor that speeds up the rate of mutation.

Practice questions

1) The grid below shows some characteristics that vary in humans.

blood group	eye colour
weight	pierced ears
sex	height

a) Identify **two** characteristics that show continuous variation.

b) Identify **one** characteristic that was not inherited.

2) The drawing below shows a pea pod in which some of the seeds are smooth and some are wrinkled. In an inheritance investigation, a pupil counted the seeds of each type in 40 pods and showed the results as a bar chart.

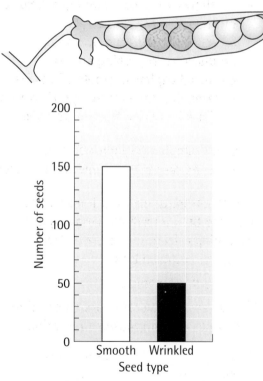

a) Is the variation in seed type continuous or discontinuous?
Give a reason for your answer.

b) Calculate the whole number ratio of smooth seeds to wrinkled.

3) The table below gives information about a cross between red-eyed and brown-eyed *Drosophila* fruit flies.

Parents	First generation	Second generation
red-eyed flies		240 red-eyed
X	all have red eyes	
brown-eyed		80 brown-eyed

a) What are the two phenotypes used in this cross?

b) Which eye colour is recessive in *Drosophila*?

c) 'P' is the shorthand symbol for parents. What is the symbol for the second generation?

d) Apart from the parents, identify a group of flies in this cross that are all true breeding.

4) The pictures below show the result of a mating between a spotted leopard and a black panther in a zoo.

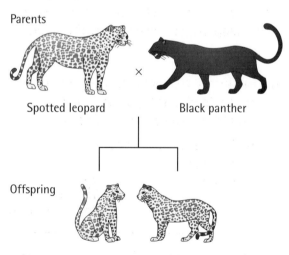

Parents

Spotted leopard   Black panther

Offspring

a) Using the symbol B for the dominant gene and b for the recessive gene give the genotypes for:
  i) the spotted parent
  ii) the black parent
  iii) the spotted offspring.

b) The letters B and b were used to represent different forms of the same gene. What word means 'the different forms of a gene'?

The picture below shows the results of mating the two cubs once they had grown.

Offspring

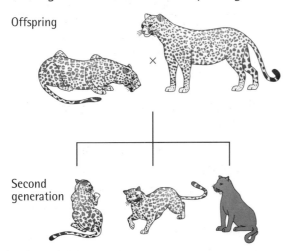

c) The zoo keepers expected a 3:1 ratio in the second generation. As you can see, they got a 2:1 ratio. Explain why the actual and predicted results were different.

d) What evidence did this cross provide to show that black panthers and spotted leopards are members of the same species?

5) The family tree below traces the inheritance of the dominant (B) short finger (brachydactyly) gene in humans.

a) What is the genotype of:
   i) Simon?
   ii) Ann?

b) What is the evidence that Moira is not true breeding for this characteristic?

6) a) Underline the correct alternative. In humans, a male has *XX/XY* sex chromosomes and a female *XX / XY*. To have a son, the mother provides an egg with the sex chromosome *X/Y* and the father a sperm with the sex chromosome *X/Y*.

b) The term used for eggs, sperm, pollen and ovules is ....

7) Decide which of the following characteristic changes are caused by mutation and which are caused by selective breeding.

a) By always planting seeds from the biggest cobs a farmer doubles the yield from his maize plot over ten years.

b) A yellow carrot was found in a field full of otherwise orange carrots.

c) A shepherd only kept ewes that produced twins, selling any that produced triplets or singles. Over his working life, the percentage of twins in his flock increased from 60 to 85%.

8) Give an example of a factor that can increase the rate at which mutations occur.

9) Human chromosome mutations can be detected before birth by removing and examining cells in the amniotic fluid surrounding the embryo.

a) Name this technique.

b) Give an example of a human condition caused by a chromosome mutation.

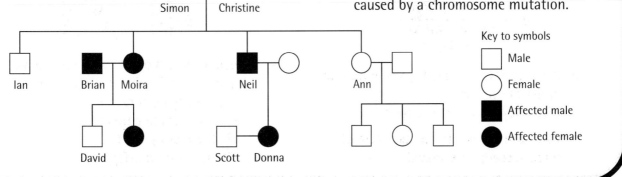

Key to symbols

☐ Male

○ Female

■ Affected male

● Affected female

Biotechnology

This topic is about:

- **living factories**
  yeast, fermentation, types of respiration, malting and brewing, cheese and yoghurt

- **problems and profit with waste**
  dangers of untreated sewage, working with micro-organisms, decay, sewage treatment, upgrading waste

- **reprogramming microbes**
  genetic engineering, insulin, antibiotics, biological detergents, immobilisation techniques

Biotechnology is the application of biological processes in industry. It takes its place at the very heart of the scientific revolution and its impact arguably rivals that of micro-electronics.

Biotechnology has a very long history. To begin, we consider the traditional products of living organisms such as wine, beer, cheese and yoghurt. The processes involved rely on various aspects of fermentation, and the differences between this and aerobic respiration are considered.

With the increase in human population, and the industrial and agricultural methods that have accompanied it, has come a huge increase in the production of waste of all types. Biotechnology has an important role in the disposal of many of these wastes, and indeed it is both possible and economically beneficial, to upgrade some of our waste to useful products such as foods and fuels using this technology.

Finally, and as an exceptionally fitting way in which to finish our study of Standard Grade Biology, we come to the reprogramming of microbes. Our ability to do this in order to allow the rapid and economical production of a range of valuable, and often life-saving, products is already in the process of revolutionising modern medicine. We consider the basics of the methods used and examples of the products that have resulted so far. The nature of biotechnology, and the incredible speed of its progress, ensures that this aspect of biology will retain its position as the unchallenged foremost science, in terms of its relevance and impact on our everyday lives, and on the planet upon which we live.

FactZONE

It would be helpful to know the meaning of some biotechnology words and phrases for the exam.

Use the following pages to find the meanings of these words and then write them beside the correct definition. Tick them off as you go.

- ☐ yeast
- ☐ fermentation
- ☐ batch processing
- ☐ malting
- ☐ lactose
- ☐ typhoid
- ☐ resistant spores
- ☐ energy
- ☐ oxygen
- ☐ biogas
- ☐ genetic engineering
- ☐ insulin
- ☐ antibiotics
- ☐ immobilisation
- ☐ continuous-flow process

Words	Meanings
	is the sugar that is found in milk
	is where all the raw materials are mixed together at the start and the product needs to be separated from the cells or enzymes at the end
	is the transfer of a gene from one species to another
	are a stage in the life cycle of bacteria or fungi when they are difficult to kill
	is the gas needed for the complete breakdown of sewage
	is a single-celled fungus
	prevent the growth of bacteria
	is where the substrate solution trickles down through immobilised enzymes and a pure product flows out
	is where barley grains are germinated so that they change their starch food store into maltose sugar
	means the same as anaerobic respiration as it takes place when oxygen is absent
	is what micro-organisms get when they decay waste and recycle elements
	a disease spread by drinking water contaminated by untreated sewage
	is a product from genetically modified bacteria that is needed by diabetics
	traps cells or enzymes on the surface of jelly beads
	is another name for the fermentation fuel methane

Living factories

Man uses the cell activities within a variety of micro-organisms to make many products of economic importance. Some produce foods, others drinks, fuels or medicines.

Yeast

Yeast is a fungus that uses sugar as food. As you can see from the picture, yeast is a single-celled organism. Given the right conditions, it multiplies rapidly by budding, which is a type of asexual reproduction.

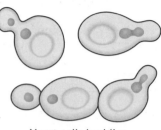

Yeast cells budding

❶ REMEMBER You may be asked to name a food or drink made by yeast.

The raising of dough and the making of beer and wine depend on the activities of yeast. Bread dough rises because of the bubbles of carbon dioxide gas given off by the yeast. Beer and wine depend on the ability of yeast to make alcohol and, in the case of beer and champagne, carbon dioxide gas for the fizz.

📺 *Watch the video sequence on the use of yeast to make wine.*

Fermentation

Fermentation is the name used to describe the release of energy from food without using oxygen. When brewers and wine makers want yeast to produce alcohol they keep it short of oxygen. The process of yeast fermenting glucose can be represented by a word equation:

$$\text{glucose} \xrightarrow{\text{yeast}} \text{alcohol + carbon dioxide + energy}$$

Types of respiration

The biological term 'respiration' describes any chemical process that releases energy from food. There are two types, depending upon whether oxygen is available or not.

- Aerobic respiration — which uses oxygen.

- Anaerobic respiration — which occurs when no oxygen is available.

❶ REMEMBER You must be able to compare aerobic and anaerobic respiration.

Aerobic respiration releases much more energy from a given mass of food. Anaerobic respiration, or fermentation, releases much less energy. Anaerobic respiration produces alcohol and carbon dioxide as by-products when it happens in plants or fungi such as yeast. Anaerobic respiration produces lactic acid as a by-product when it happens in animals and bacteria.

 Underline the correct alternative.
Fermentation occurs in the presence/absence of oxygen.

Optimum conditions for brewing

Commercial brewers make sure that beer production is as efficient as possible by providing the best possible conditions for yeast to grow and ferment. This means that the temperature, oxygen supply and concentration of glucose must be carefully controlled. Also, unwanted micro-organisms that might spoil the quality or flavour of the beer must be kept out.

As the diagram shows, the yeast in the fermentation vessel can be warmed or cooled to regulate the temperature.

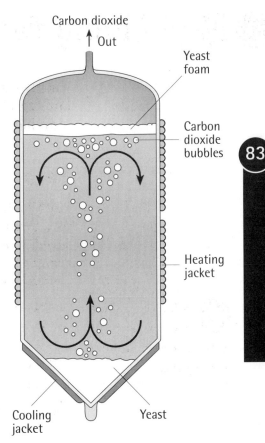

Carbon dioxide
↑ Out

Yeast foam

Carbon dioxide bubbles

Heating jacket

Cooling jacket

Yeast

Batch processing

The usual way brewers make sure that the optimum conditions are present throughout the brewing process is to carefully set up the fermentation vat at the start. Yeast, malt, hops, sugar and water are mixed together and then the whole system is left closed, but controlled, until the fermentation is complete. The vat must be thoroughly cleaned between batches. Also, the yeast needs to be filtered out of the beer at the end. This has the disadvantage that production is only possible in batches rather than as a continuous process. A production process like this is called 'batch processing'.

Malting

The brewing industry uses barley as the source of food that the yeast ferments to make the alcohol in beer. However, barley stores food in the form of starch — which is a type of food that yeast cannot use. In order to solve this problem, the maltster allows the barley seeds, or grains, to germinate. The enzyme amylase in the barley seed converts its own starch into maltose sugar, which yeast can ferment. This process is called malting.

> **REMEMBER**
> You must be able to describe how brewers provide the best growing conditions for yeast.

Milk products — cheese and yoghurt

Making cheese and yoghurt are also processes that depend upon micro-organisms — in this case, bacteria. The souring of milk is also a fermentation process. Fresh milk contains some bacteria, which feed on the sugars in the milk. The main sugar in milk is called lactose, which is converted into lactic acid by bacterial fermentation. The process of bacteria fermenting lactose sugar can be represented by the following word equation:

$$\text{lactose sugar} \xrightarrow{\text{bacteria}} \text{lactic acid} + \text{energy}$$

The increased acidity, or lower pH of the lactic acid, makes the milk turn sour and curdle.

📺 *Watch the video sequence about making cheese using bacteria.*

❓ *Which type of organism is used to make cheese and yoghurt?*

> **REMEMBER**
> In the exam, you may be asked to name the type of sugar found in milk.

Problems and profit with waste

Environmental damage from untreated sewage

The disposal of untreated sewage causes damage to water environments. For example, it lowers the oxygen concentration of rivers, which makes it impossible for fish and many invertebrates to survive. Untreated sewage can also spread diseases caused by harmful micro-organisms. Such diseases include typhoid, food poisoning, polio, dysentery and cholera.

(?) *The removal of oxygen from rivers results in .... conditions.*

Safety precautions with micro-organisms

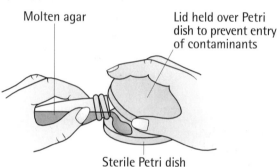

Molten agar

Lid held over Petri dish to prevent entry of contaminants

Sterile Petri dish

It is important to have scrupulously clean conditions when working with micro-organisms in the laboratory. Hands need to be thoroughly washed before and after the experiment, benches have to be washed down with a disinfectant, and all equipment must be sterilised. Careful techniques must be used when transferring micro-organisms to prevent cross contamination. The microbes must be killed by a high temperature once the experiment is over.

Similar precautions are needed in biotechnology industries to avoid the contamination of pure cultures of microbes by other unwanted micro-organisms. A particular danger in these cases are spores of bacteria and fungi that are resistant to normal hygiene measures. Very high temperatures and strong chemicals are used to kill them. Careful and regular checks for contamination by stray bacteria and fungi are made at all stages of manufacture.

Decay

Decay is both important and useful because it recycles raw materials and gets rid of waste at the same time. Decay is the decomposition of organic matter by bacteria and fungi. The micro-organisms feed on the organic matter to provide themselves with energy. During the decay process, both carbon and nitrogen are recycled.

! REMEMBER Note that typhoid is a disease that is commonly spread by sewage-contaminated drinking water.

Sewage treatment

In a sewage-treatment works, organic waste is broken down by the action of decay micro-organisms to products that are less harmful to the environment. The micro-organisms need lots of oxygen to do their job properly. As the picture on page 85 shows, this is provided by jets of compressed air, or by stirring the waste mechanically.

Plentiful supplies of oxygen mean that the bacteria use aerobic respiration, which allows all the molecules in the sewage to be completely broken down. Anaerobic respiration by micro-organisms would only partly break down the waste.

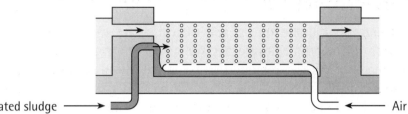

Recycled activated sludge ⟶   ⟵ Air

Sewage contains a wide variety of waste materials. Different types of micro-organism feed on different waste materials. This means that a range of different micro-organisms must be present so that all the waste can be broken down completely.

📺 *Watch the video sequence showing the workings of a sewage treatment plant.*

❗ **REMEMBER** Aerobic conditions are needed for the complete breakdown of sewage.

Upgrading waste

The main aim in upgrading waste is to convert it to more useful substances. Most commonly this involves raising the levels of protein, or increasing the energy content of the waste. This technology is important because environmentally damaging substances are converted into substances such as food and fuel.

Food from micro-organisms

Under suitable conditions, micro-organisms can reproduce very rapidly. These micro-organisms can be grown on organic waste such as whey from cheese making, or fruit pulps after juice extraction. The micro-organisms can then be harvested to provide protein-rich food for animals or man.

Fermentation fuels

Alcohol and methane are both valuable fuels. Each of them can be produced by micro-organisms. Biogas, which is mainly methane, is produced by fermenting organic waste such as sewage sludge and domestic refuse. Alcohol is obtained by fermenting suitable plant crops such as sugar cane. The production of fuel in this way has certain advantages over the use of fossil fuels such as coal and oil. Fermentation fuels use renewable materials as opposed to fossil fuels, which will eventually run out. In many cases, it allows countries with no coal or oil to produce their own fuel. The raw materials used are often the waste from other production processes, which could otherwise be harmful or difficult to dispose of. Burning fermentation fuels produces less atmospheric pollution than burning fossil fuels.

❗ **REMEMBER** You should be able to state that upgrading waste raises the level of protein or energy content.

Genetic engineering

The control of all the normal activities of a bacterium depends upon its single circular chromosome. It also has a few genes in smaller plasmids. In genetic engineering, a gene from a different organism can be inserted into the plasmid. This allows the engineered bacteria to make a substance that bacteria have never made before. The picture shows the steps in the genetic engineering process to produce bacteria able to make insulin.

REMEMBER You should be able to explain how a human gene is inserted into a bacterium using a plasmid.

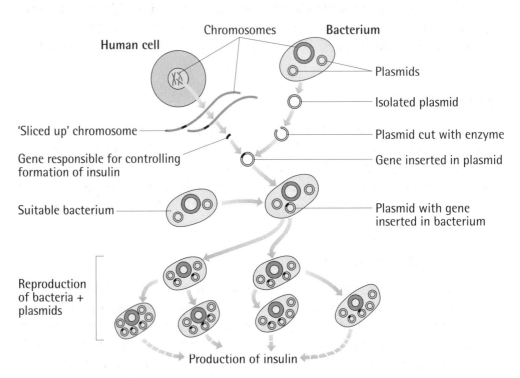

As a result of genetic engineering, it is possible to use bacteria to produce unlimited quantities of a product. Genetic engineering also speeds up the process of producing many new substances.

REMEMBER You should know that genetic engineering produces the desired organism much faster than selective breeding.

 Watch the video sequence on insulin and the genetic engineering process.

Genetic engineering vs selective breeding

Genetic engineering is a way of producing organisms with new genotypes best suited for a particular function. In the past, humans have used selective breeding to achieve this.

Genetic engineering has several advantages over selective breeding as a method. Single characteristics can be modified instead of dealing with the complicated genetics of plants or animals. A desirable characteristic that can make the product faster can be transferred from one species into another. The process can be accomplished in a few years instead of over many generations.

Genetic engineering is used for the production of substances that used to be difficult to produce. Examples include insulin for the control of diabetes, some antibiotics and various vaccines for the control of disease.

Insulin

The need for insulin is increasing for two reasons. Diabetes often doesn't occur until old age and better treatment means many more patients are living longer. In addition, the human population is increasing.

Animal insulin from slaughtered cattle and pigs was not as effective as human insulin, and some patients were allergic to it. The insulin produced by genetically modified bacteria is pure human insulin and so it has none of the problems associated with the use of animal insulin.

Antibiotics

An antibiotic is a chemical that prevents the growth of micro-organisms. There are many diseases, each caused by a different bacteria. The diagram (right) shows that the growth of the bacteria is prevented by streptomycin; but penicillin is ineffective. So, one antibiotic may only work against certain types of bacteria. This means that a range of antibiotics is needed for the treatment of bacterial diseases.

Bacterial colonies

P   S

Penicillin disc   Streptomycin disc

Biological detergents

Biological detergents contain enzymes produced by bacteria. Many of the stains on clothes, such as sweat, blood and grass, are proteins. The bacterial enzymes break down and digest proteins remove the stains. The enzymes work at relatively low temperatures to remove stains, which would otherwise need high temperature washes. This saves energy by allowing low temperature washes. It also helps to clean delicate fabrics, which would be damaged by a very hot wash.

! REMEMBER
Antibiotics prevent the growth of bacteria.

Immobilisation techniques

Immobilisation techniques restrict the movement of enzymes or cells. This is usually done by attaching them to beads of jelly or other surfaces. Immobilisation allows continuous flow processing by effectively keeping the enzymes or cells separate from the product.

Unlike batch processing where fresh enzyme needs to be used for each batch, the continuous-flow process allows the enzyme to be used for a long period of time. Also, the product is pure and so further expensive processing to separate the product from the enzyme is not necessary.

! REMEMBER
You need to know that biological detergents contain enzymes that are produced by bacteria.

 Check your understanding with the Test bite.

Practice questions

1) Underline the correct alternative.

   Bread dough rises because of the carbon dioxide gas released by *yeast/bacteria*. Making cheese relies on *yeast/bacteria* to sour and curdle the milk. Wine and beer producers use *yeast/bacteria* to ferment sugars to alcohol and carbon dioxide. In addition, *yeast/bacteria* are used by the dairy industry to manufacture yoghurt.

2) The graph below shows the growth of an antibiotic-producing fungus.

   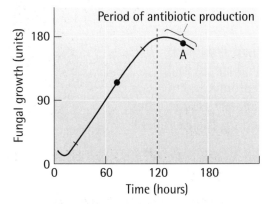

   a) Fungi require suitable conditions for growth. State how **two** such conditions are provided.

   b) How many hours after the culture was set up did the fungus start producing penicillin?

   c) State one factor that could have been limiting the growth of the fungal population at A.

3) Underline the correct alternative to make this sentence correct.

   The *fungus/bacterium* yeast is a *multi-/single* -celled organism that uses *sugar/starch* as a source of food.

4) Complete this word equation of the fermentation of sugar by yeast.

   $$\text{..............} \xrightarrow{\text{yeast}} \text{..............} + \text{carbon dioxide}$$

5) Decide which of the following statements refer to aerobic and which to anaerobic respiration of glucose:

   a) oxygen essential
   b) lactic acid produced
   c) glucose fully broken down, releasing all its energy
   d) water is a byproduct
   e) alcohol and carbon dioxide released
   f) occurs in the absence of oxygen
   g) also known as fermentation.

6) Complete the following sentences.

   Malting involves providing .... seeds with suitable conditions for germination. The seeds produce an enzyme called .... , which breaks down the starch in the food store to .... sugar. Malting is required before brewing as .... can use sugar for food but not starch.

7) The manufacture of cheese and yoghurt begins with fermentation by micro-organisms.

   a) Name the type of micro-organism used in cheese and yoghurt making.

   b) What is the name of the sugar in milk?

   c) Which product of milk sugar fermentation causes it to sour and curdle?

8) a) Describe **two** precautions used during safe laboratory work with micro-organisms.

   b) Which is the resistant stage of the life cycle of bacteria when they are most difficult to kill?

9) It is important for the health of the environment and man for sewage water to be treated before being released into a river.

   a) Describe an example of environmental damage caused by the discharge of untreated sewage into a river.

b) Give an example of a disease spread by untreated sewage.

c) Describe how the oxygen required for complete breakdown is provided in the sewage treatment works.

d) Explain why a range of micro-organisms is needed in sewage treatment.

e) What do the micro-organisms obtain as repayment for breaking down the sewage for us?

f) Give an example of an important element that is recycled by sewage decay.

10) The following steps of the genetic engineering process are not in the correct order.

i) multiply the bacteria, which contain the insulin gene
ii) cut out the insulin gene
iii) obtain plasmids from bacteria and cut them open
iv) collect and purify the human insulin made by the bacteria
v) replace the plasmids in bacteria
vi) insert the insulin genes into the plasmids
vii) identify the insulin gene in a human chromosome.

Rewrite the steps in the correct sequence.

11) a) Give **two** reasons to explain why the need for insulin is increasing each year.

b) Describe the advantage of genetically engineered insulin over insulin produced from slaughtered mammals.

c) Give an example of another economically important product from genetic engineering.

d) Explain why genetic engineering is regarded as being more efficient than selective breeding at producing new genotypes.

12) Biological detergents are now used frequently in Scottish homes.

a) What kind of biologically active ingredient do they use to digest stains?

b) Give an advantage of using a biological detergent instead of a non-biological detergent.

13) The diagram below shows the effect of antibiotics on a Petri dish of bacteria.

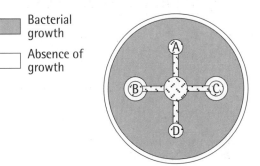

Bacterial growth

Absence of growth

a) What effect do antibiotics have on bacteria?

b) Which antibiotics had an effect on these bacteria?

c) How do the results of this experiment suggest that a range of antibiotics are required?

14) Decide which of the following statements refer to the batch process and which to the continuous-flow process.

a) The vat is cleaned out before refilling.

b) Immobilized enzymes are used.

c) All the ingredients are mixed together at the start.

d) A pure product flows out.

e) The substrate is added gradually.

f) An extra process is needed to separate the product from the catalyst.

Answers to questions

The biosphere (page 18)

1 The place where an organism lives is called a _habitat_. All the animals and plants that live in an area are known as a _community_. The organisms of one species that live in an area are known as a _population_. All the plants, animals and abiotic factors in an area are called an _ecosystem._

2 a) The stone casts out light, which many invertebrates avoid, stops the rain getting in and the birds eating your catch.

   b) Beetles have eaten everything else.

   c) More traps and visit them often.

3 Add them up and divide the total by the number of samples. So, 5 + 7 + 10 + 3 + 0 + 3 + 12 + 5 + 6 + 9 = 60. Then, 60/10 = 6.

4 Meters are used to measure abiotic factors. A _light_ meter is used to measure light _intensity_. The meter must not be _shaded_ with the body as this would give a reduced reading. When using a moisture meter experimental error is reduced by _wiping_ the probe between samples. The _reliability_ of the results from both meters can be increased by taking several readings.

5 a) The picture shows a king salmon.

   b) Chum has marks above the lateral line and silver on both sides.

6 a) A producer is a green plant that makes its own food by photosynthesis.

   b) There are two food chains with five organisms in the web. They are: leaves → beetles → spiders → hedgehogs → foxes; AND bark → woodlice → spiders → hedgehogs → foxes.

   c) The arrows represent the direction of energy flow.

   d) The squirrel eats acorns and the eggs and chicks of blackbirds, so it is an omnivore.

   e) i) The spider population will increase because it is no longer being eaten by the hedgehogs.

      ii) Several answers are reasonable here. The population of blackbirds could increase as there are more worms for them to eat. OR The population of blackbirds might decrease as the hungry foxes might eat more of them. OR The population of blackbirds may stay the same as more food is balanced by more predation.

   f) Both eat leaves.

   g) Energy is lost from this food web by movement and heat production.

7 A pyramid of _biomass_ shows the _weight_ of organisms in each stage of a food chain.

8 a)

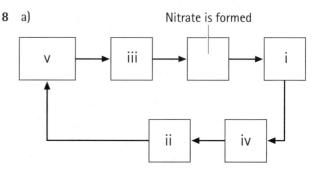

   b) Micro-organisms obtain energy when they decay waste.

9 a) i) Agricultural pollutants include pesticides, fertilisers and slurry.

      ii) Industrial pollutants include oil, sulphur dioxide and heavy metals.

      iii) Domestic pollutants include detergents, sewage and household waste.

   b) The smoke from burning fossil fuels contains sulphur dioxide, which leads to acid rain, and carbon dioxide, which adds to global warming.

10 a) Sewage is an organic waste.

   b) The bacteria in the river use the organic waste as food and multiply.

   c) As the number of bacteria increase, the concentration of oxygen decreases.

   d) The smallest number of species are likely to be 2 km downstream where the oxygen concentration is lowest.

11 An indicator species is an organism that shows the level of pollution by being present or absent.

World of plants (page 28)

1 a) Decrease in plant variety means a reduced variety of food, raw materials and medicines for man.

   b) Animals lose their food and habitats, and so animal species become extinct.

2 A = raw material, B = food, C = medicine.

3 The brewing industry uses barley as a source of food for yeast. The maltster encourages the barley grains to germinate. The enzyme amylase in the barley grain converts its own starch into maltose sugar. This is malting.

4 In the future, plants may provide man with new foods, raw materials or medicines.

5 a) i) Seed coat protects the seed.
    ii) Embryo grows into a plant.
    iii) Food store provides energy for germination and early growth.
  b) Water, oxygen and a suitable temperature are required for germination.

6 a) i) petal = C, ii) stigma = A, iii) anther = B.
  b) i) The stigma is feathery and hangs out of the flower to filter pollen from the wind.
    ii) The anther hangs out of the flower to release pollen into the wind.
  c) Pollination is the transfer of pollen from anther to stigma.
  d) Coloured petals attract insects.

7 Soon after the pollen grain lands on the _stigma_ it germinates and begins to grow a pollen _tube_. This grows down into the ovule inside the _ovary_. The pollen _nucleus_ then flows down to fuse with the ovule _nucleus_. This fusion is known as fertilisation.

8 A = animal internal, B = wind, C = animal external.

9 a) Potato tubers and strawberry runners are examples of asexual reproduction.
  b) Advantages of asexual reproduction include passing on good characteristics and growing in a clump to reduce competition from other plants.
  c) A clone is a group of genetically identical plants, e.g. a field of Golden Wonder potato plants.
  d) Advantages of sexual reproduction include variation in the offspring and the spread of the offspring to new areas.

10 a) i) xylem, ii) phloem, iii) phloem, iv) phloem, v) xylem.
  b) Xylem supports the plant.

11 a) B = epidermis, E = stoma, I = xylem.
  b) i) guard cells = F.
    ii) most photosynthesis occurs in C, which is the palisade mesophyll.
  c) Part A, the cuticle, makes the leaf waterproof, cutting down water loss.
  d) Carbon dioxide diffuses in.

  e) Water vapour and oxygen diffuse out through the stoma.

12 carbon dioxide + water $\xrightarrow[\text{chlorophyll}]{\text{light}}$ food + oxygen

13 Chlorophyll converts _light_ energy to _chemical_ energy. Using this energy, carbon dioxide is converted to the storage carbohydrate known as _starch_ and the structural carbohydrate called _cellulose_.

Animal survival (page 42)

1 All food types contain carbon, hydrogen and oxygen. Only _protein_ contains nitrogen. Simple sugar molecules joined together form _carbohydrates_, and amino acids linked up make _proteins_. A _fat_ molecule is formed when fatty acids and glycerol combine.

2 Digestion breaks down large insoluble molecules into smaller insoluble molecules.

3 a) A = incisor.
  b) The dagger-like canines pierce and hold the prey animal.
  c) A carnivore eats other animals.
  d) i) Your example of a herbivore should be a plant-eater such as a goat.
    ii) Your example of an omnivore should be a plant- and meat-eater such as a human, pig or bear.

4
Starch $\xrightarrow{\text{Amylase}}$ Maltose sugar

Protein $\xrightarrow{\text{Pepsin}}$ Peptides

Fat $\xrightarrow{\text{Lipase}}$ Fatty acids + glycerol

5 a) i) pancreas = C, ii) appendix = G, iii) gall bladder = I, iv) oesophagus = A.
  b) i) stores bile = I, ii) makes pepsin = B, iii) absorbs food = D.

6 Sperm have a _tail_ for swimming. Eggs are much larger than sperm as they contain more _food_. In fish, fertilisation is _external_; whereas fertilisation is _internal_ in mammals to provide _water_ for the sperm to swim in.

7 The more protection provided by the parents during fertilisation and development of the embryo, the greater the chance of survival and so fewer eggs need to be produced. Some fish give no protection and so must release millions of eggs. Some fish give quite a lot of protection and the eggs needed are in the hundreds. Mammals provide a lot of protection, even after birth, and so need to make few eggs.

BITESIZEbiology

8 a) A = vagina, B = uterus, C = oviduct, D = ovary.
   b) i) eggs are made = D, ii) fertilisation occurs = C,
       iii) fetus develops = B.

9 i) A fish embryo gets its energy for growth and
       development from the yolk in the egg.
   ii) A mammal embryo obtains its energy from its
       mother via the placenta.

10 A mammal loses water in urine, faeces, sweat and in
    exhaled air. It gains water in food, by drinking and
    from aerobic respiration.

11 a) C = ureter.
    b) D is the bladder, which stores urine.
    c) Blood vessel A is the renal artery because it has an
        arrow going into the kidney.

12 a) Urea is made from excess amino acids.
    b) Urea is made in the liver.
    c) Urea is transported to the kidneys in the blood.

13 Sweating *lowers* the water concentration of the
    blood. The brain detects the imbalance and releases
    more ADH. When the ADH arrives at the kidney it
    causes an *increase* in water reabsorption. As a result a
    smaller volume of urine is produced.

14
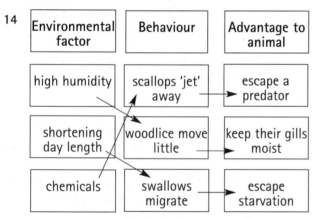

Environmental factor	Behaviour	Advantage to animal
high humidity	scallops 'jet' away	escape a predator
shortening day length	woodlice move little	keep their gills moist
chemicals	swallows migrate	escape starvation

Investigating cells (page 52)

1 *Cells* are the basic units of living things.

2 a) i) cell wall = B, ii) cytoplasm = A, iii) large central
        vacuole = C.
   b) i) The nucleus controls all the cell activities
           including cell division.
       ii) The chloroplast is the site of photosynthesis.
       iii) The cell membrane controls which molecules
           enter and leave a cell. It is semi-permeable so
           only small molecules can pass through.
   c) Animal cells contain only a nucleus, cytoplasm and
       membrane.

3 Stains are used to make certain parts of a cell more
   visible. For example, iodine makes the transparent
   nucleus orange.

4 a) Molecules move by diffusion from a *high*
       concentration to a *lower* concentration *down* a
       concentration gradient.
   b) Really, a cell could be found to suit any answer
       you have here as long as the molecules are small.
       Pick any two from water, oxygen, carbon dioxide,
       urea and amino acids.
   c) Oxygen diffuses into cells so energy can be
       released.
   d) The membrane controls diffusion in and out
       of cells.

5 Osmosis is the movement of *water* molecules from a
   high water concentration to a *lower water*
   concentration through a *semi-permeable* membrane.

6 a) Cell A has been in tap water.
   b) Water moved out of cell B.
   c) The cells were left for 30 minutes to allow osmosis
       time to have an effect.
   d) Lettuce leaves go limp when they lose water and
       would have cells like B.

7 When blood cells are placed in a strong salt solution
   the cells *shrink* because water moves *out* from a *high*
   water concentration to a *lower* water concentration.

8 a) The chip would become shorter, lighter and more
       flexible.
   b) The reliability of the results could have been
       improved by repeating the experiment with
       several chips.

9 a) The word that means the same as cell division is
       mitosis. This is the preferred term in exams.
   b) The correct order is E, D, C, A then B.
   c) The chromatids are pulled to the poles of the cell.
       Remember to use the word chromatid, or you
       won't get any marks.
   d) So that the daughter cell has the full set of
       information.
   e) The nucleus controls mitosis, as well as all the
       other cell activities.

10 This shows that the liver contained an *enzyme* which
    acted as a biological *catalyst* because it increased the
    rate of a chemical reaction. A control for this
    experiment would have everything exactly the same
    except there would be no *liver*.

11 a) Several possibilities from amylase, pepsin, lipase
        and catalase.

b) Phosphorylase is a synthesis enzyme.

c) Protein.

12 a) As the temperature rises, the activity of amylase increases until 50°C. As the temperature increases further, the activity decreases.

b) Amylase is most active at its optimum temperature.

c) Enzymes are specific, so pepsin could not digest starch.

13 Fat has more energy per gram than protein or carbohydrate.

14 The most important product released during aerobic respiration is _energy_. For food to be broken down completely _oxygen_ gas must be present. Water and _carbon dioxide_ gas are also made.

Body in action (page 66)

1 a) i) The skull protects the brain.
ii) The rib cage protects the heart and lungs.
iii) The vertebrae protect the spinal cord.

b) Two other functions of the skeleton are support and a framework for muscle attachment.

2 a) The elbow is a hinge joint, which allows movement in only one direction.

b) Tendons attach muscles to bone.

c) Muscle T contracts and muscle B relaxes to straighten the arm.

3 The protein fibres in bone, which gave flexibility, have been burnt away leaving the hard, but brittle, mineral part.

4 People with an energy input that is _less_ than their energy output lose weight.

5 a) A = bronchus, B = bronchiole, C = air sac.

b) Cartilage rings keep the trachea open.

c) When the diaphragm contracts, it moves down, which increases the volume of the chest. This decreases the pressure in the chest and air rushes into the lungs.

d) Dust and bacteria stick to the mucus. Cilia sweep the dirty mucus up and out of the lungs.

6 a) B = left atrium, E = aorta.

b) F is the pulmonary vein, which returns blood from the lungs.

c) Chamber D, the left ventricle, has thick muscular walls as it has to pump blood all the way around the body; whereas chamber C, the thinner-walled right ventricle, just has to pump blood to the lungs.

d) Structure V is a valve. Like all valves, it stops the backflow of blood. In this case from the L ventricle into the L atrium.

e) The heart is provided with its own blood supply by the coronary artery.

7 a) vein, b) capillary, c) artery, d) artery, e) capillary, f) vein.

8 a) Oxygen is the only molecule carried by red blood cells. Many soluble molecules are carried in the plasma such as carbon dioxide, glucose, amino acids, urea, salt and proteins.

b) Haemoglobin is the oxygen-carrying pigment in red blood cells.

9 a) A = iris, B = lens, C = pupil.

b) The retina converts light energy to electrical energy, and the optic nerve carries electrical impulses to the brain.

c) Two eyes allow the brain to judge distance.

10 The fluid-filled canals are in three different planes so head movement in any direction can be detected.

11 a) A = cerebrum, B = cerebellum.

b) The medulla controls the heart and breathing rate.

12 a) Fatigue.

b) The word equation for anaerobic respiration is:
food → lactic acid + energy.

13 a) 80 beats per minute.

b) Recovery time means the period between stopping exercise and the heart rate returning to its resting value.

c) Isobel is fitter than Morag because her heart rate did not rise as high during exercise and her recovery time was shorter.

Inheritance (page 78)

1 a) The two characteristics in the grid that show continuous variation are weight and height.

b) The characteristic that was not inherited is pierced ears. This has nothing to do with the genes passed on by parents.

2 a) Variation in seed type is discontinuous variation because smooth and wrinkled seeds form two distinct groups with nothing in between.

b) Divide the larger number of 150 by the smaller number of 50 to get **3**. So, the ratio is 3:1.

3 a) The phenotypes are the appearances of the organisms and so, in this cross, they are red eye and brown eye.

b) All the first generation and the majority of the second generation are red-eyed so that is the dominant gene. The recessive gene is brown eye.

c) F_2 is the symbol used for second generation.

d) Remember, true breeding means that two genes are the same. Two dominant or two recessive. The brown eyes in the F_2 are certainly true breeding as they must have two recessive genes.

4 a) i) the spotted parent is BB, ii) the black parent is bb, iii) the spotted offspring have a gene from both parents so they will be Bb.

b) The word 'allele' means different forms of the same gene.

c) A good answer to explain a difference between expected and actual results is that 'fertilisation is random'. Also the size of the sample – three cubs – is small. We might get nearer to a 3:1 ratio if we looked at several litters from this pair.

d) Surprisingly, despite different names and appearances, the spotted leopard and black panther are members of the same species. The cross shows that they were able to produce offspring and, more importantly, their cubs had cubs of their own.

5 a) i) Simon must have a dominant (B) gene as he has the condition. He must also have a recessive (b) gene to pass on to Ian and Ann. So he is Bb.
   ii) Ann is bb as she does not have short fingers.

b) Remember, true breeding means that both genes are the same. Moira and her husband Brian both have the condition. If either of them was a true breeding BB then their son David could not be bb and would have full-length fingers.

6 a) In humans, a male has _XY_ sex chromosomes and a female _XX_. To have a son the mother provides an egg with the sex chromosome _X_ and the father a sperm with the sex chromosome _Y_.

b) The term used for eggs, sperm, pollen and ovules is _gametes_.

7 a) = selective breeding, b) = mutation, c) = selective breeding.

8 Factors that increase the rate of mutations include chemicals, high temperatures and radiation, such as X-rays and ultraviolet light.

9 a) Amniocentesis is a way of detecting chromosome mutations before birth.

b) Down's syndrome is a human condition caused by a chromosome mutation.

Biotechnology (page 88)

1 Bread dough rises because of the carbon dioxide gas released by _yeast_. Making cheese relies on _bacteria_ to sour and curdle the milk. Wine and beer producers use _yeast_ to ferment sugars to alcohol and carbon dioxide. In addition, _bacteria_ are used by the dairy industry to manufacture yoghurt.

2 a) A suitable temperature, oxygen and a food supply.

b) 120 hours.

c) Lack of food or a build up of waste products.

3 The _fungus_ yeast is a _single_-celled organism that uses _sugar_ as a source of food.

4 $$\text{sugar} \xrightarrow{\text{yeast}} \text{alcohol} + \text{carbon dioxide}$$

5 a) aerobic, b) anaerobic, c) aerobic, d) aerobic, e) anaerobic, f) anaerobic, g) anaerobic.

6 Malting involves providing _barley_ seeds with suitable conditions for germination. The seeds produce an enzyme called _amylase,_ which breaks down the starch in the food store to _maltose_ sugar. Malting is required before brewing as _yeast_ can use sugar for food but not starch.

7 a) Bacteria are used to make cheese and yoghurt.

b) Milk sugar is known as lactose.

c) Lactic acid sours and curdles milk.

8 a) Any two from: hands must be thoroughly washed, benches wiped down with disinfectant, sterile equipment used, aseptic techniques for microbe transfer, and all organisms autoclaved at the end.

b) The spores of bacteria and fungi are especially resistant.

9 a) Untreated sewage provides food for the river bacteria causing them to multiply. The billions of bacteria use very little oxygen each but, together, they use up all the oxygen in that part of the river. Other organisms die or move away because of the oxygen shortage. This means that the number of species in that part of the river is reduced.

b) Typhoid and cholera.

c) Oxygen is added by stirring or by injecting compressed air.

d) Sewage consists of a wide range of materials, which requires a broad spectrum of micro-organisms to digest it completely.

e) Micro-organisms obtain energy from the decay of waste.

f) Nitrogen and carbon are recycled during the decomposition of sewage.

10

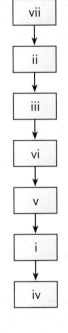

11 a) The need for insulin is increasing as the world population is increasing and people are living longer.

b) Genetically engineered insulin has an unlimited supply and does not cause allergic reactions.

c) Other genetically engineered products include vaccines, enzymes and antibiotics.

d) Genetic engineering is a faster way of changing the genotype of organisms.

12 a) Biological detergents contain enzymes produced by bacteria.

b) Biological detergents work at low temperatures and so require less energy per wash. In addition, delicate fabrics can be cleaned at low temperatures without being damaged.

13 a) Antibiotics prevent the growth of bacteria.

b) Antibiotics B and C prevented the growth of this particular bacteria.

c) Only some antibiotics are effective against a particular bacteria.

14 a) batch, b) continuous, c) batch, d) continuous, e) continuous, f) batch

Index